Mind and Brain

Vida Demarin

Editor

Mind and Brain

Bridging Neurology and Psychiatry

 Springer

Editor
Vida Demarin
International Institute for Brain Health
Zagreb, Croatia

ISBN 978-3-030-38605-4   ISBN 978-3-030-38606-1   (eBook)
https://doi.org/10.1007/978-3-030-38606-1

This Springer imprint is published by the registered company Springer Nature Switzerland AG
The registered company address is: Gewerbestrasse 11, 6330 Cham, Switzerland

Introduction

The brain as an organ and mind as its product and even more than that were for centuries a secret that many scientists tried to reveal. But only recently, during and following Decade of the Brain, with the introduction of neuroimaging methods, especially functional magnetic resonance imaging (fMRI), it was possible to get more data on these important topics. This was a kind of a "golden era" for neurology and psychiatry, disciplines that got tools for deeper searching and possibilities of getting answers to important questions, together with results from neuroscience and neurobiology.

This development enabled getting a broader field of the investigation, the interaction of several disciplines, new perspectives, and a more comprehensive view of mind and brain interaction.

Our Mind & Brain Congress was founded 60 years ago as a small gathering of colleagues with a special interest in brain diseases. From that time onward, it was intensively developing—always being international, interdisciplinary, focused at recent knowledge, spreading it among participants from all over the world, on the path of what was the founding idea of "Pula School of Science and Humanity".

And today, with emerging a new discipline Psychoneuroendocrinoimmunology and its translation to clinical practice, our long-standing comprehensive approach to the interrelation and bridging of the mind and the brain could contribute to a better understanding of this challenging topic. As it is not possible to show in detail this huge area of interest, we have tried with a few examples in the following chapters to illustrate it from different perspectives.

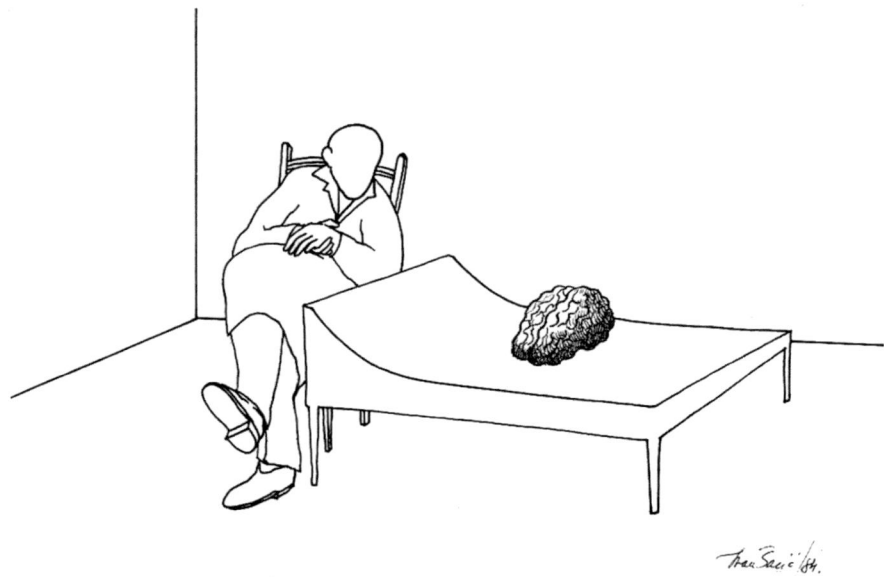

Brain on the couch (author: Ivan Šarić, courtesy from the private collection)

Contents

Contributors

Goran Babić Croatian Psychoanalytic Society, Zagreb, Croatia

Karl Bechter Department Psychiatry and Psychotherapy 2, University of Ulm, Ulm, Germany

Natan Bornstein Brain Division, Shaare Zedek Medical Center, Jerusalem, Israel

Hrvoje Budinčević Department of Neurology, Sveti Duh University Hospital, Zagreb, Croatia;
Faculty of Medicine, Josip Juraj Strossmayer University of Osijek, Osijek, Croatia

Petra Črnac Žuna Department of Neurology, Sveti Duh University Hospital, Zagreb, Croatia

Miroslav Cuturic Department of Neurology, University of South Carolina School of Medicine, Columbia, SC, USA

Vida Demarin International Institute for Brain Health, Zagreb, Croatia

Filip Derke Department of Neurology, Clinical Hospital Dubrava, Zagreb, Croatia

Anton Glasnović Croatian Institute for Brain Research, Zagreb, Croatia

Milija Mijajlović Clinical Center of Serbia and School of Medicine University of Belgrade, Neurology Clinic, Belgrade, Serbia

Sandra Morović Juraj Dobrila University of Pula, Pula, Croatia

Igor Mošič NAM Emotional Hygiene Technology, Rijeka, Croatia

Osman Sinanović School of Medicine University of Tuzla, Tuzla, Bosnia and Herzegovina;
School of Science and Technology, Sarajevo Medical School University Sarajevo, Sarajevo, Bosnia and Herzegovina

Tena Sučić Radovanović Department of Radiology, Sveti Duh University Hospital, Zagreb, Croatia

Vladimira Vuletić Clinical Department of Neurology, University Hospital Centre Rijeka, Rijeka, Croatia;
Medical Faculty University of Rijeka, Rijeka, Croatia

Marjan Zaletel Clinical Department of Vascular Neurology, University Medical Centre Ljubljana, Ljubljana, Slovenia

Bojana Žvan Clinical Department of Vascular Neurology, University Medical Centre Ljubljana, Ljubljana, Slovenia

Creativity—The Story Continues: An Overview of Thoughts on Creativity

Vida Demarin and Filip Derke

What is Creativity? Where does it come from? Are only 'special' people creative and 'ordinary' are not? Are we born with it? Or can it be learned, improved and enhanced?

These and similar questions are being present for many years. People were always curious to get the right answer. Modern neuroscience with sophisticated neuroimaging methods has already given many answers to this important topic and intensive research is still going on.

Creativity is the act of turning new and imaginative ideas into reality.

It can perceive the world in new ways, to find hidden patterns, to make connections between seemingly unrelated phenomena and to generate solutions.

It involves two processes: thinking and producing.

Creativity is a process of bringing something new into being, it requires passion and commitment. It brings to our awareness what was previously hidden and points to new life. The experience is one of heightened consciousness: ecstasy.—Rollo May, The Courage to Create [1].

According to Albert Einstein, Creativity is intelligence having fun, while for Henri Matisse, Creativity takes Courage.

The human brain supports different kinds of creativity but different human brains lead to different kinds of creativity. It is nowadays present in many, almost all segments of our life, from classrooms to workplaces, science, artistic expression, to home decoration, to business, marketing, recreation, even to stylish food preparation.

Beliefs that only special, talented people are creative (and one has to be born that way) diminish our confidence in our creative abilities.

V. Demarin (✉)
International Institute for Brain Health, Ul. Grada Vukovara 271, HR-10000 Zagreb, Croatia
e-mail: vida.demarin@gmail.com

F. Derke
Department of Neurology, Clinical Hospital Dubrava, Av. Gojka Šuška 6, HR-10000 Zagreb, Croatia

© Springer Nature Switzerland AG 2020
V. Demarin (ed.), *Mind and Brain*, https://doi.org/10.1007/978-3-030-38606-1_1

The notion that geniuses such as Shakespeare, Picasso and Mozart were 'gifted' is a myth, according to a study at Exeter University. Researchers examined outstanding performances in the arts, mathematics and sports, to find out if 'the widespread belief that to reach high levels of ability a person must possess an innate potential called talent'. The study concludes that excellence is determined by opportunities, encouragement, training, motivation, and most of all, practice [2].

But special attention was always given to artistic talent. Orator fit, poeta nascitur—'The orator is made; the poet is born'—is an ancient and well-known expression and today it is being supported more and more often by scientific evidence. Many scientists have been interested in the particularities of the ways in which artists think, work and create, encouraging them to do research through which they would attempt to explain why artists are more special than most other 'ordinary' people and whether, by means of modern technology, it is possible to get insight to the structure and function of the brain and it is possible to discover the secret of—or rather, 'embody' the quality that we call talent.

Csikszentmihalyi [3], after more than 30 years of observing creative people, wrote:

'If I had to express in one word what makes their personalities different from others, it is complexity'.

They show tendencies of thought and action that in most people are segregated. They contain contradictory extremes; instead of being an 'individual,' each of them is a 'multitude.'

Can Creativity Be Taught?

George Land's Creativity Test [4]

In 1968, George Land conducted a research study to test the creativity of 1,600 children ranging from 3 to 5 years of age who were enrolled in a Head Start program. This was the same creativity test he devised for NASA to help select innovative engineers and scientists. The assessment worked so well that he decided to try it on children. He retested the same on children at 10 years of age, and again at 15 years of age. The results were astounding. Among 5-year-olds, creativity was present in 98%. When they were retested on 10-year-olds, it was found in 30%. After 5 more years, when they were 15 years old, the result was even worse. Creativity was present in only 12% of them. The same test was given to 280,000 adults and creativity was present in only 2% of them.

The conclusion from this study is obvious: Non-creative behaviour is learned. It is the following results obtained at Exeter university, encouragement, training, motivation, and most of all, practice are the most important for further development and boosting of this innate feature.

Fostering creativity in children in all domains and directions of their interest is 'a must' for parents, teachers and everybody taking care of children and their development. For nurturing a child's imagination and creativity, it is good to spend time outdoors. The benefits of nature on a child's development are endless. Because nature is ever-changing, it provides countless opportunities for discovery, creativity and problem-solving. As Albert Einstein said: Imagination is more important than knowledge, it is of utmost importance to use our imagination in enhancing the children's one.

Where Creativity Comes from? Is There a Special Place in the Brain?

Modern neuroscience has the privilege to investigate the processes of artistic performance in a healthy brain using modern techniques such as functional MRI (magnetic resonance imaging). Not so long-ago scientists could only speculate what brain functions are involved in artistic processes by observing neurological patients [5]. In the process of explaining the secret of creativity, a simplified theory started with the known fact of cerebral hemisphere dominance; uncreative people have marked hemispheric dominance and creative people have less marked hemispheric dominance [6, 7]. The right hemisphere is specialized, among other functions, for metaphoric thinking, for playfulness, solution-finding and synthesizing. It is the centre of visualization, imagination and conceptualization, but the left hemisphere is still needed for artistic work to achieve balance by partly suppressing creative states of the right hemisphere and for the executive part of a creative process. Numerous studies investigating the brain function during the visual art activities have shown a very specific functional organization of brain areas. Different parts of visual cortex were activated, depending on the type of picture viewed (colours, objects, faces, the position of objects in space, motion or static pictures) [8]. Marked hemispheric dominance and area specialization is very prominent for music perception. Both brain hemispheres are needed for complete music experience, while frontal cortex has a significant role in rhythm and melody perception. The centres for perceiving pitch and certain aspects of melody and harmony and rhythm are identified in the right hemisphere. The left hemisphere is important for processing rapid changes in frequency and intensity of tune. Several brain imaging studies have reported activation of many other cortical areas besides auditory cortex during listening to music, which can explain the impact of listening to music on emotions, cognitive and motor processes [9, 10].

Drawing 2: Brain Tree (Author: Ivan Šarić, Courtesy from the Private Collection)

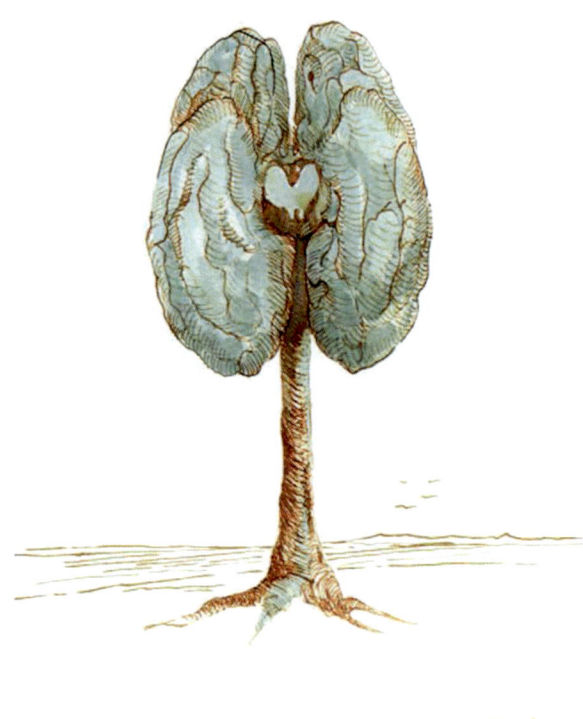

Where it comes from, is crucial for some product to become a piece of art: the creativity arising from an artist's brain is necessary. But it is also interesting to establish why a great number of people find a particular piece of art, music, dance or a poem beautiful. The saying 'The beauty is in the eye of the beholder' is known from ancient times. Recently, Tomohiro Ishizu and Semir Zeki conducted a study the results of which have revealed that a beauty experience is indeed in the beholder, though not in the eye, but the brain [11]. As well as for the art experience, Zeki pointed out in his book 'A Vision of the Brain' that all human experience is mediated through the brain and is not solely the product of the outside world. He says, 'The more important the experience, the more it can reveal about the fundamental properties of the brain' [12].

Contrary to the 'right-brain' myth, which was considered as a rule for many years, now we are witnessing data that creativity does not just involve a single brain region or even a single side of the brain. Instead, the creative process draws on the whole brain. It is a dynamic interplay of many different brain regions, emotions and our unconscious and conscious processing systems.

In an attempt to explain this process, three most frequently used networks are as follows [13, 14]:

Network 1: The Executive Attention Network—it is activated when intensive attention is needed, with heavy demands at working memory.

It is situated in the lateral region of the prefrontal cortex (PFC) and areas towards the back.

Network 2: The Imagination Network (Default Network)

'Constructing dynamic mental simulations based on personal past experiences such as used during remembering, thinking about the future and generally when imagining alternative perspectives and scenarios to the present'. The Imagination Network is also involved in social cognition—deep insula, PFC and temporal lobe communicating with parietal lobe.

Network 3: The Salience Network—Constantly monitor both external and internal stream of consciousness, what is most salient for a particular task—dorsal anterior cingulate cortices (2dACC) and anterior insula (AI).

Creativity might not be a 'right-brain' thing after all—but rather the ability of both hemispheres to communicate. A new study from Duke University [15] reveals that highly creative people have significantly more nerve connections between the right and left hemispheres than less creative people.

For the study, healthy college-age volunteers underwent MRI imaging using diffusion tensor imaging, which enables following the brain's white matter tracts by tracing the movement of water along with them.

The white matter connections of 68 different brain regions were analyzed by statisticians David Dunson of Duke University and Daniele Durante of the University of Padova.

Computers Produced 3-D Wiring Diagrams of the Brain

Neuroscientist Rex Jung of the University of New Mexico collected the MRI data and performed a variety of tests to assess participants' level of creativity. Some involved problem-solving tests that evaluated divergent thinking, such as coming up with as many users as possible for an object like a paperclip or a brick. Another example was seeing how many geometric shapes participants could draw in 5 min. Jung's team also administered a survey that asked about achievements in ten areas including creative writing, cooking, science and music. In combination, the results created a composite creativity score for each participant. Then a computer program examined the data from the brain maps and identified structural differences.

The results revealed that participants who were in the top 15% based on their creativity score had significantly more connections between the right and left hemispheres of the brain compared to those in the bottom 15%.

Most of the differences were seen in the frontal lobe, which is essentially the 'control panel' of personality and the ability to communicate. The researches did not find any statistical differences in the number of connections between men or women or within each hemisphere.

'Maybe by scanning a person's brain we could tell what they are likely to be good at', Dunson said in a statement. Noting that the technique might be able to predict how likely it is that a person will be creative based on their brain network structure.

Dunson's team is now investigating whether brain connectivity is different depending on I.Q. and is also using their approach to help distinguish early Alzheimer's disease from normal ageing.

But nevertheless, there is some grain of truth to the left-brain/right-brain distinction. For instance, spatial reasoning recruits more structures in the right hemisphere, and language processing recruits more structures in the left hemisphere. Also, there is some really interesting research conducted by Kounios and Beeman [16] showing that the A-ha moment of insight—in which participants discover seemingly unrelated words—is associated with activation of the right anterior superior temporal gyrus. None of these findings, however, negate the fact that the entire creative process involves the whole brain [17].

Despite advances in the neuroscience of creativity, the field still lacks clarity on whether a specific neural architecture distinguishes the highly creative brain. Using methods in network neuroscience, in their recent study, Beaty and colleagues modelled individual creative thinking ability as a function of variation in whole-brain functional connectivity. They identified a brain network associated with creative ability comprised of regions within default, salience and executive systems—neural circuits that often work in opposition. Across four independent datasets, they show that a person's capacity to generate original ideas can be reliably predicted from the strength of functional connectivity within this network, indicating that creative thinking ability is characterized by a distinct brain connectivity profile [18].

'Creativity is complex, and we're only scratching the surface here, so there's much more work that's needed', says Beaty.

Expertise and training in fine motor skills have been associated with changes in brain structure, function and connectivity. Fewer studies have explored the neural effects of athletic activities that do not seem to rely on precise fine motor control (e.g., distance running). In the recent study [19], resting-state functional connectivity in a sample of adult male collegiate distance runners ($n = 11$; age $= 21.3 \pm 2.5$) was compared to a group of healthy age-matched non-athlete male controls ($n = 11$; age $= 20.6 \pm 1.1$), to test the hypothesis that expertise in sustained aerobic motor behaviours affects resting-state functional connectivity in young adults. Although generally considered an automated repetitive task, locomotion, especially at an elite level, likely engages multiple cognitive actions including planning, inhibition, monitoring, attentional switching and multitasking and motor control. Connectivity was examined in three resting-state networks that link such executive functions with motor control: the default mode network (DMN), the frontoparietal network (FPN) and the motor network (MN). Two key patterns of significant between-group differences in connectivity that are consistent with the hypothesized cognitive demands of elite endurance running were found. First, enhanced connectivity between the FPN and brain regions often associated with aspects of working memory and other executive functions (frontal cortex) suggest endurance running may stress executive cognitive functions in ways that increase connectivity in associated networks. Second, there was also found significant anti-correlations between the DMN and regions associated with motor control (paracentral area), somatosensory functions (postcentral region) and visual association abilities (occipital cortex). DMN deactivation with task-positive regions is generally beneficial for cognitive performance, suggesting anti-correlated regions observed here are engaged during running. For all between-group differences, there were associations between connectivity, self-reported physical activity and estimates of maximum aerobic capacity, suggesting a dose-response relationship between engagement in endurance running and connectivity strength. Together these results suggest that differences in experience with endurance running are associated with differences in functional brain connectivity. High-intensity aerobic activity that requires sustained, repetitive locomotor and navigational skills may stress cognitive domains in ways that lead to altered brain connectivity, which in turn has implications for understanding the beneficial role of exercise for the brain and cognitive function over the lifespan.

This latest University of Arizona study concludes that similar brain changes are gained via the precise, fine-tuned motor skills associated with playing a musical instrument and the seemingly repetitive action of sustained aerobic motor behaviours, such as endurance running.

Future studies will focus on pinpointing how exercise-induced or musical training-induced neuroplasticity in young adults might improve cognitive function in the short and long-term. Regular physical activity and musical training seem to share the ability to promote greater brain connectivity and better cognitive function across the human lifespan.

Creativity is a vast construct, seemingly intractable to scientific inquiry—perhaps due to the vague concepts applied to the field of research. One attempt to limit the purview of creative cognition formulates the construct in terms of evolutionary constraints, namely that of blind variation and selective retention (BVSR). Behaviourally, one can limit the 'blind variation' component to idea generation tests as manifested by measures of divergent thinking. The 'selective retention' component can be represented by measures of convergent thinking, as represented by measures of remote associates [20]. The authors summarize results from measures of creative cognition, correlated with structural neuroimaging measures including structural magnetic resonance imaging (sMRI), diffusion tensor imaging (DTI), and proton magnetic resonance spectroscopy (1H-MRS). They also review lesion studies, considered to be the 'gold standard' of brain–behavioural studies. What emerges is a picture consistent with theories of disinhibitory brain features subserving creative cognition, as described previously [21]. They provide a perspective, involving aspects of the default mode network (DMN), which might provide a 'first approximation' regarding how creative cognition might map on to the human brain.

Their review suggests that when you want to loosen your associations, allow your mind to roam free, imagine new possibilities and silence the inner critic, it's good to reduce activation of the Executive Attention Network (a bit, but not completely) and increase activation of the Imagination and Salience Networks. Indeed, recent research on jazz musicians and rappers engaging in creative improvisation suggests that is precisely what is happening in the brain while in a flow state.

In a recent review, the authors outlined how three factors crucially shape the creative mind: (1) creative cognition and the associated neural systems in human and animal models; (2) creative drives such as mood states, emotion, motivation and regulatory focus and how their interactions could shape the creative performance; and (3) the impacts of three central neuromodulatory systems, i.e., dopaminergic (DA), noradrenergic (NE), and serotonergic (5-HT), on the interplay between creative cognition and creative drives [22].

Philosophers and scientists have long puzzled over where human imagination comes from. In other words, what makes humans able to create art, invent tools, think scientifically and perform other incredibly diverse behaviours?

Drawing 3: Flower Brain (Author: Ivan Šarić, Courtesy from the Private Collection)

The answer, Dartmouth researchers conclude in a new study, lies in a widespread neural network—the brain's 'mental workspace'—that consciously manipulates images, symbols, ideas and theories and gives humans the laser-like mental focus needed to solve complex problems and come up with new ideas.

Eleven areas of the brain are showing differential activity levels in a Dartmouth study using functional MRI to measure how humans manipulate mental imagery [23].

Another fMRI study investigated brain activation during creative idea generation using a novel approach allowing spontaneous self-paced generation and expression of ideas, trying to get the answer to the fundamental question of what brain processes are relevant for the generation of genuinely new creative ideas, in contrast to the mere recollection of old ideas from memory. In general, creative idea generation (i.e., divergent thinking) was associated with extended activations in the left prefrontal cortex and the right medial temporal lobe, and with the deactivation of the right temporoparietal junction. The generation of new ideas, as opposed to the retrieval of old ideas, was associated with stronger activation in the left inferior parietal cortex which is known to be involved in mental simulation, imagining and future thought. Moreover, brain activation in the orbital part of the inferior frontal gyrus was found to increase as a function of the creativity (i.e., originality and appropriateness) of ideas pointing to the role of executive processes for overcoming dominant but uncreative responses. Authors conclude that the process of idea generation can be generally understood as a state of focused internally directed attention involving controlled semantic retrieval. Moreover, left inferior parietal cortex and left prefrontal regions may subserve the flexible integration of previous knowledge for the construction of new and creative ideas [24].

The importance of the Default mode network in creativity was investigated in a structural MRI study. A positive correlation was found between creative performance and grey matter volume of the default mode network. These findings support the idea that the default mode network is important in creativity and provide neurostructural support for the idea that unconscious forms of information processing are important in creativity [25].

A novel game-like and creativity-conducive fMRI paradigm is developed to assess the neural correlates of spontaneous improvisation and figural creativity in healthy adults. Participants were engaged in the word-guessing game of Pictionary™, using an MR-safe drawing tablet and no explicit instructions to be 'creative'. Using the primary contrast of drawing a given word versus drawing a control word (zigzag), authors observed increased engagement of cerebellum, thalamus, left parietal cortex, right superior frontal, left prefrontal and paracingulate/cingulate regions, such that activation in the cingulate and left prefrontal cortices negatively influenced task performance. Further, using parametric fMRI analysis, increasing subjective difficulty ratings for drawing the word engaged higher activations in the left prefrontal cortices, whereas higher expert-rated creative content in the drawings was associated with increased engagement of bilateral cerebellum. Altogether, these data suggest, rather unexpected result, that cerebral–cerebellar interaction underlying implicit processing of mental representations has a facilitative effect on spontaneous improvisation and figural creativity [26].

In his excellent review on the possible mechanism of creativity, Heilman [27] points out that creative people are often risk-takers and novelty seekers behaviours that activate their ventral striatal reward system. Innovation also requires associative and convergent thinking, activities that are dependent on the integration of highly distributed networks. People are often most creative when they are in mental states associated with reduced levels of brain norepinephrine, which may enhance the communication between distributed networks.

Based on this assumption, creative arts-based therapies are used in poststroke rehabilitation and enhancing the quality of life. Different art modalities are perceived to be useful in achieving different therapeutic goals. Interventions that offer opportunities for the participants to experience different art modalities during the process may foster participation and enhance flexibility. Therefore, further research is needed to demonstrate the differential benefits or special advantages in using single or multiple art modalities as well as having qualified therapists in creative arts-based therapies [28, 29].

Years ago, Ellis Paul Torrance has started investigation in the field of creativity, especially in children [30–34]. He pointed out the role of the teacher, the importance of developing children's creativity in the classroom and other scientific views.

A great amount of research related to the assessment of creativity was done by Robert Sternberg and his colleagues [35], who established the so-called Investment Theory of Creativity and Development 1991 [36].

Music and Neuroplasticity

There is tremendous research in the field of music, which is proven to be very important in our life. Music is a unique stimulus that activates almost every cognitive system in the brain. It activates so many parts of our brain that it is impossible to say that we have a centre for music the way we do for other tasks and subjects, such as language. When we hear a song, our frontal lobe and temporal lobe begin processing the sounds, with different brain cells working to decipher things like rhythm, pitch and melody. In his book This is Your Brain on Music, The Science of Human Obsession, Levitin discusses these data in detail [37].

Music is a strong stimulus for neuroplasticity. fMRI studies have shown reorganization of the motor and auditory cortex in professional musicians. Other studies showed the changes in neurotransmitter and hormone serum levels in correlation to music [38].

Brain scans reveal that listening to pleasing music increases activity in the brain's reward centre and triggers the release of dopamine. Scientists believe that music's ability to make you feel good maybe one of the ways it helps alleviate the pain. The more pleasant the listeners found the music to be, the less pain they felt, what was thoroughly explained in Oliver Sacks's brilliant book Musicophilia [39].

Music can interfere with pain signals even before they reach the brain—at the level of the spinal cord and eventually can prevent the transmission of pain signals to the brain [37, 39].

The very prominent connection between music and enhancement of performance or changing of neuropsychological activity was shown by studies involving Mozart's music from which the theory of 'The Mozart Effect' was derived [40].

Brain images of people listening to music show activity in different parts of the brain [41] and there are particular areas of the brain which show idiosyncrasies in musicians including the motor cortex, the cerebellum and the corpus callosum.

Results of other numerous studies showed that listening to music can improve cognition, motor skills and recovery after brain injury.

To investigate the neural substrates that underlie spontaneous musical performance, [42] authors examined improvisation in professional jazz pianists using functional MRI. By employing two paradigms that differed widely in musical complexity, they found that improvisation (compared to the production of over-learned musical sequences) was consistently characterized by a dissociated pattern of activity in the prefrontal cortex: extensive deactivation of the dorsolateral prefrontal and lateral orbital regions with focal activation of the medial prefrontal (frontal polar) cortex. Such a pattern may reflect a combination of psychological processes required for spontaneous improvisation, in which internally motivated, stimulus-independent behaviours unfold in the absence of central processes that typically mediate self-monitoring and conscious volitional control of ongoing performance. Changes in prefrontal activity during improvisation were accompanied by widespread activation of neocortical sensorimotor areas (that mediate the organization and execution of musical performance) as well as deactivation of limbic structures (that regulate motivation and emotional tone). This distributed neural pattern may provide a cognitive context that enables the emergence of spontaneous creative activity.

Music, according to old Chinese saying—music has the power to ease tension within the heart and lessen and loosen obscure emotions.

The connection between the brain and music is strong and bidirectional. Like Oliver Sacks, a professor of neurology and a writer who extensively studied the effect of music on human health wrote: 'We turn to the music, we need it, because of its ability to move us, to induce feelings and moods, states of mind' [39].

Types of Creativity and Creative Thinking

Creativity is often confined and even mixed up with artistic talent, a lot of dispute is still going on. Because of this, creativity is often assumed as something lateral, without focusing what is inside it. We should not just discount it. We should think of creativity as our intelligence with two facets: crystallized and fluid, what can be also said for creativity. It is a bit of both and neither at the same time. Dietrich [43] has identified four different types of creativity with corresponding brain activities—a kind of matrix.

Creativity can either be emotionally (based from the heart) or cognitively (thoughts, logical intelligence) based, and can also be spontaneous (something unexpected) or deliberate (conscious effort to sustain), what gives us four quadrants, each with different and unique aspects to it!

Deliberate and Cognitive—Thomas Edison

Deliberate and cognitive creativity comes from sustained work in the discipline and previous body of knowledge. The prefrontal cortex (PFC)—pays focused attention and makes connections among the information that have been stored—putting together existing information in new and novel work. This type of creativity requires a high degree of knowledge and plenty of time and patience.

Personal Breakthrough 'A-ha' Moments

Deliberate, emotional–personal 'A-ha' moment is characterized by our vibrant spectrum of emotion and its deliberate nature to make us feel, this can be encountered whenever suddenly happens an insight about ourself. Also, PFC is activated and deliberate part and cingulate cortex that processes feelings that are related, how we interact with others and what is our place in the world.

Isaac Newton 'Eureka' Moments

Working on the idea or a problem that we can't solve need break for getting insight, after which we can work better. Basal ganglia, where dopamine is stored are involved. When this type of creativity happens, the conscious brain stops working on the problem and hands it over to the unconscious brain. By doing this, our brains can connect new information to the task via our unconscious mental processes. By doing different, unrelated activities, PFC can connect the information in new ways. Existing knowledge belongs to the cognitive part.

'Epiphanies'

Due to the nature of our emotions, epiphanies are difficult to find. This is spontaneous and emotional creativity coming from the amygdala—where basic emotions are processed. When conscious brain and PFC are resting, spontaneous ideas and creations emerge. It happens to artists and musicians, and it is quite a powerful feeling, such as epiphany. It is not cognitive, no knowledge, but skill is necessary—writing, artistic, musical …

Deliberate and cognitive creativity requires a high degree of knowledge and lots of time, deliberate and emotional creativity requires quiet time, spontaneous and cognitive creativity requires stopping work on the problem and getting away while spontaneous and emotional creativity probably cannot be designed for.

After years of research, investigating types of thinking, Ned Herrmann developed his concept of Whole-Brain Thinking System [44] with four types of thinking, each corresponding roughly to one of brain structure. This concept is combining the left and right theory and Paul MacLean's triune theory [45] in a new model of thinking styles. Each person has a preference in his/her's thinking style and has to be aware that another person has also his thinking style. Four patterns that emerged in terms of how the brain perceives and processes information have led to this Herrmann's concept and he constructed a measuring tool for seeing the difference, called Herrmann Brain Dominance Instrument (HBDI).

In his book The Creative Brain [46], Ned Herrmann explains his four specialized clusters of mental activity: A/Facts—Goals driven: Logical, Analytical, Fact-based, Quantitative intelligence. B/Form—Results Driven: Structural–Operational, Sequential, Organized, Detailed Planned Intelligence. C/Feelings—People Driven: Social–Relational Intelligence, Interpersonal, Feeling Based, Emotional Intelligence and D/Future—Vision Driven: Conceptual Creative, Intuitive, Integrating, Syntethesizing Intelligence.

Creativity is personal, we should claim for our creative space at work, within the family, everywhere and unleash our enormous creative capability.

Apart from Herrmann's ways of thinking, convergent and divergent thinking are also described as a possible mode of creative thinking. They are the two ways of approaching problem-solving [47]. Divergent thinking seeks multiple perspectives and multiple possible answers to questions and problems while convergent thinking assumes that a question has only one right answer and that a problem has a single solution [48]. The divergent thinker does not accept the usual way of doing things, to be his way, he always tries to get more ideas and more ways of doing it. Divergent thinking is the process of thought where a person uses flexibility, fluency and originality to explore as many solutions or options to a problem or issue as possible. It is the opposite of convergent thinking, which has the characteristic to focus on only one idea or single solution, being rather conservative, always accepting the answer as being the right, and not even trying to seek for other options. For divergent thinkers, it is rather easy to find additional ideas and answers, contrary to convergent thinkers who are not that good in finding new solutions. But, on the other hand, without convergent thinking, it would be rather difficult to come to an end and to conclude, as Guilford [49] wrote years ago, and research is going on [50].

Researching through a vast amount of literature on creative thinking, Jorlen [51] pointed out to five main types of creative thinking: Divergent thinking (exaggeration), Lateral thinking (out-of-the-box), Aesthetic thinking (Beauty and taste), Systems thinking (Synthesis toward a whole) and Inspirational thinking (Emergent, radical insight).

These five types of creative thinking are the basis, but much of the research around creativity paints a more nuanced picture.

There are similarities and correlations between the five clusters, and instead of five, there could be even ten. Another challenge is, that much of the literature on creativity is based on personal stories, which makes any classification difficult as it can never be proven whether different accounts of creativity and creative thinking are referring to the same experiences for different people.

Divergent thinking is already mentioned, and brainstorming is a typical example of divergent thinking, where 'downloading' or emptying the brain of a certain topic takes place. This technique is however limited in that it builds on releasing the ideas that are already stored in a person's brain, and not to generate any new ideas.

Lateral Thinking

Creativity involves breaking out of established patterns to look at things differently—Edward De Bono [52].

Creativity researcher Edward De Bono came up with the term 'lateral thinking' in 1967 to 'distinguish between artistic creativity and idea creativity'. The term was invented as an alternative to step-by-step thinking, so-called vertical thinking, which is justified with sequential steps based on logic. Lateral thinking can be used for the generation of new ideas and problem-solving as it by definition leaves the already used behind and looks for completely new options. This type of thinking is based on avoiding the intrinsic limitations in the brain, which rapidly sees patterns and handles information in a distinctive way, where long thought sequences are not broken up once formed. Instead, lateral thinking tools and techniques can be used to restructure and escape such as 'clichéd' patterns and think 'outside the box'.

Aesthetic Thinking

It took me four years to paint like Raphael, but a lifetime to paint like a child—Pablo Picasso

This type of thinking involves producing or discovering things, which are pleasant, harmonious and beautiful to our senses. It is an ancient form of thinking which can be learned by anyone.

It is, however, important to emphasize that this type of creative thinking might be enough to build a story but to create a great work of art, other types of creative thinking are needed too. The same goes for all work, which is built on aesthetic thinking. A person will not become a great artist only by going to art school.

Systems Thinking

> Creativity is just connecting things. When you ask creative people how they did something, they feel a little guilty because they didn't do it, they just saw something. It seems obvious to them after a while Steve Jobs once said.

Systems thinking can be described as the ability to see how things are interrelated and form a larger 'whole'. Some people seem to be able to perceive such links more easily than others, to 'connect the dots' and understand that if one thing is changed, the whole system will change.

It is closely related to aesthetic thinking in that synthesis and making things 'whole' and perfect, somehow is related to elegance and beauty. It is also closely related to the next type of thinking—inspirational thinking.

Inspirational Thinking

This type of creative thinking concerns the perception of receiving insights from somewhere or someone else. It often happens in dreams or other states, but sometimes in extremely powerful, rapid bursts of clarity and focus, known as light-bulb moments or peak experiences.

As already discussed, by the use of neuroscience methods, it is now possible to document what happens in the brain during the creative process. MRI—functional MRI and especially diffusion tensor imaging for evaluation of connectivity is of utmost importance in collecting new data and building the broader and more exact picture. But the results from research in the field of neurophysiology, especially EEG and evoked potentials during the last approximately 20 years are also contributing to getting more insight on the neural basis of creativity.

The focus was mainly given towards changes in alpha power related to types of thinking and various task solving, being frequently present in divergent thinking, in creative idea production, during active thinking and focused attention. Meditation has been found as a possible method to enhance creativity [53, 54]. The interrelation between intelligence and creativity should be considered in a closer look [55].

Reviewing the emerging literature on the neural basis of creativity, Dietrich and Kanso [56] classified studies into three categories: divergent thinking, artistic creativity and insight.

Electroencephalographic studies of divergent thinking yield highly variegated results. Neuroimaging studies of this paradigm also indicate no reliable changes above and beyond diffuse prefrontal activation. These findings call into question the usefulness of the divergent thinking construct in the search for the neural basis of creativity. A similarly inconclusive picture emerges for studies of artistic performance, except that this paradigm also often yields activation of motor and temporoparietal regions. Neuroelectric and imaging studies of insight are more consistent, reflecting changes in anterior cingulate cortex and prefrontal areas. Taken together, creative thinking does not appear to critically depend on any single mental process or brain region, and it is not especially associated with right brains, defocused attention, low arousal or alpha synchronization, as sometimes hypothesized. To make creativity tractable in the brain, it must be further subdivided into different types that can be meaningfully associated with specific neurocognitive processes.

In investigating task-related alpha power changes during creative activity, the observed time course of alpha activity may reflect the progression of different stages in the process of idea generation: an initial bilateral alpha synchronization followed by a relative decrease in alpha power and an increasing hemispheric lateralization driven by a reincrease of alpha power at right frontal and posterior cortical sites [57].

Investigating the different patterns of EEG alpha activity related to convergent and divergent task processing, divergent processing was found to involve higher task-related EEG alpha power than convergent processing in both the alternate uses task and the word association task. EEG alpha synchronization can hence explicitly be associated with divergent cognitive processing rather than with general task characteristics of creative ideation tasks. Further results point to a differential involvement of frontal and parietal cortical areas by individuals of lower versus higher trait creativity [58].

This distinctive patterns of task-related alpha activity as a function of time reflect the sequence of well-known stages of the creative idea generation process: initial retrieval of common and old ideas, followed by the actual generation of novel and more creative ideas by overcoming typical responses through processes of mental stimulation and imagination.

EEG alpha activity is sensitive to different creativity-related demands. Creativity is associated with alpha increases at frontal and right parietal sites. Alpha increases during creative cognition reflect internal processing demands [59].

Cortical activity in the human electroencephalogram alpha band was measured (using an event-related approach) in a pre- and a post-test (with intermediate training) while participants ($n = 30$) were confronted with divergent thinking tasks. Half of the participants received a divergent thinking training (over 2 weeks) which was composed of exercises structurally similar to those used in the pre- and post-test. Analyses revealed that the training group displayed higher task-related synchronization of frontal alpha activity (i.e., increases in alpha power from the prestimulus reference to the activation interval) than the control group. These findings are in line with the view of frontal alpha synchronization as a selective top-down inhibition process that prevents internal or top-down information processing being disturbed by incoming external input [60].

Cortical activity in the EEG alpha band has proven to be particularly sensitive to creativity-related demands, but its functional meaning in the context of creative cognition has not been clarified yet. Specifically, increases in alpha activity (i.e., alpha synchronization) in response to creative thinking can be interpreted in different ways: As a functional correlate of cortical idling, as a sign of internal top-down activity or, more specifically, as selective inhibition of brain regions. We measured brain activity during creative thinking in two studies employing different neurophysiologic measurement methods (EEG and fMRI). In both studies, participants worked on four verbal tasks differentially drawing on creative idea generation. The EEG study revealed that the generation of original ideas was associated with alpha synchronization in frontal brain regions and with a diffuse and widespread pattern of alpha synchronization over parietal cortical regions. The fMRI study revealed that task performance was associated with strong activation in frontal regions of the left hemisphere. Besides, we found task-specific effects in parietal–temporal brain areas. The findings suggest that EEG alpha band synchronization during creative thinking can be interpreted as a sign of active cognitive processes rather than cortical idling [61].

Another investigation done by the same group from Graz showed that Alpha power increase in right parietal cortex reflects focused internal attention [62].

Intelligence and creativity are usually studied as two separate cognitive faculties assessed with standard problems having well- or ill-defined problem spaces; intelligence primarily focuses on finding the correct solution, creativity in generating new approaches. The view emerges, however, that they play complementary roles and may be more related than research recognizes. In the present study, participants ($N = 52$) created their intelligence tasks: 3×3 matrices featuring relations between geometrical components. Using task-related alpha synchronization, it was demonstrated that intelligence integrates with creativity in a problem-solving process evolving in open problem space. The activity was especially visible at prefrontal and frontal sites when information processing was most demanding, i.e., at the start of the creative process. This research could open the way to an approach to cognition where intelligence-related abilities are studied due to multiplicity of ideas, and at the end due to narrowing down alternatives in open problem spaces [49].

In several books written by experts in the field of creativity, Arne Dietrich [63], Nancy Andreasen [64], Oshin Vartanian [17], Robert Keith Sawyer [65], Ivan Šarić [66], Eric Kandel [67], Elkhonen Goldberg [68], Anna Abraham [69] and more, different views, understanding and approaches could be found.

In spite of huge research on creativity that has been, and still is going on, the best would be to conclude with Eric Jerome Dickey's thought: It's impossible to explain creativity, It's like asking a bird, 'How do you fly?' You just do!

References

1. May RR. The courage to create. New York: Norton; 1975.
2. Renner J, Bailie B. Wisdom keys for releasing your creative potential. CreateSpace Independent Publishing Platform; 2012.

3. Csikszentmihalyi M. The systems model of creativity: the collected works of Mihaly Csikszentmihalyi. Dordrecht: Springer; 2014. https://doi.org/10.1007/978-94-017-9085-7.
4. Land G, Jarman B. breaking point and beyond. San Francisco: Harper Business; 1993.
5. Demarin V. et al. Arts, brain and cognition. Psych Danubina. 2016;28(4):343–8.
6. Katz AN. Creativity and the right cerebral hemisphere: towards a physiologically based theory of creativity. J Creat Behav. 1978;12:253–64. https://doi.org/10.1002/j.2162 6057.1978.tb00173.x.
7. York GK: The cerebral localization of creativity. In: Rose FC, editor. Neurology of the arts: paintings, music, literature. London: Imperial College Press; 2004. p. 1–9. https://doi.org/10.1142/9781860945915_0001.
8. Zeki S. Neural concept formation and art: Dante, Michaelangelo, Wagner. In: Rose FC, editor. Neurology of the arts: paintings, music, literature. London: Imperial College Press; 2004. p. 13–41. https://doi.org/10.1142/9781860945915_0002.
9. Tramo MJ, Shah GD, Braida LD. Functional role of auditory cortex in frequency processing and pitch perception. J Neurophysiol. 2002;87:122–39. https://doi.org/10.1152/jn.00104.1999.
10. Janata P, Grafton ST. Swinging in the brain: shared neural substrates for behaviors related to sequencing and music. Nat Neurosci. 2003;6:682–7. https://doi.org/10.1038/nn1081.
11. Ishizu T, Zeki S. Toward a brain-based theory of beauty. PLoS One. 2011;6(7):e21852. https://doi.org/10.1371/journal.pone.0021852.
12. Zeki S. A vision of the brain. Oxford: Blackwell Scientific Publications; 1993.
13. De Pisapia N, Bacci F, Parrott D, Melcher D. Brain networks for visual creativity: a functional connectivity study of planning a visual artwork. Sci Rep. 2016. https://doi.org/10.1038/srep39185.
14. Beaty RE, Benedek M, Wilkins RW, et al. Creativity and the default network: a functional connectivity analysis of the creative brain at rest. Neuropsychologia. 2014;64:92–8. https://doi.org/10.1016/j.neuropsychologia.2014.09.019.
15. Durante D, Dunson DB. Bayesian inference and testing of group differences in brain networks. Bayesian Anal. 2017. https://doi.org/10.1214/16-BA1030.
16. Kounios J, Beeman M. The A-ha! moment: the cognitive neuroscience of insight. Curr Dir Psychol Sci. 2009;18(4):210–6. https://doi.org/10.1111/j.1467-8721.2009.01638.x.
17. Vartanian O, Bristol AS, Kuaufman JC. Neuroscience of creativity. MIT Press; 2016.
18. Beaty RE, et al. Robust prediction of individual creative ability from brain functional connectivity. PNAS. 2018. https://doi.org/10.1073/pnas.1713532115.
19. Limb CJ, Braun AR. neural substrates of spontaneous musical performance: an fMRI study of Jazz improvisation. Greene E, editor. PLoS One. 2008;3(2):e1679. https://doi.org/10.1371/journal.pone.0001679.
20. Jung RE, Mead BS, Carrasco J, Flores RA. The structure of creative cognition in the human brain. Front Hum Neurosci. 2013;7:330. https://doi.org/10.3389/fnhum.2013.00330.
21. Martindale C. Creativity and primary process thinking. Contemp Psychol. 1981;26:568. https://doi.org/10.1037/020404.
22. Khalil R, et al. The link between creativity, cognition and creative drives and underlying neural mechanisms. Front Neural Circuits. 2019. https://doi.org/10.3389/fncir.2019.00018.
23. Schlegel A, Kohler PJ, Fogelson SV, Alexander P, Konuthula D, Tse PU. Network structure and dynamics of the mental workspace. Proc Natl Acad Sci USA. 2013;110(40):16277–82. https://doi.org/10.1073/pnas.1311149110.
24. Benedek M, Jauk E, Fink A, et al. To create or to recall? Neural mechanisms underlying the generation of creative new ideas. Neuroimage. 2014;88(100):125–33. https://doi.org/10.1016/j.neuroimage.2013.11.021.
25. Kühn S, Ritter SM, Müller BCN, van Baaren RB, Brass M, Dijksterhuis A. The importance of the default mode network in creativity-a structural MRI study. J Creat Behav. 2014;48:152–63. https://doi.org/10.1002/jocb.45.

26. Saggar M, Quintin E-M, Kienitz E, et al. Pictionary-based fMRI paradigm to study the neural correlates of spontaneous improvisation and figural creativity. Sci Rep. 2015;5:10894. https://doi.org/10.1038/srep10894.

27. Heilman KM. Possible brain mechanisms of creativity. Arch Clin Neuropsychol. 2016;4:285–96. https://doi.org/10.1093/arclin/acw009.

28. Lo TLT et al. Creative arts-based therapies for stroke survivors, a qualitative systematic review. Front Psychol. 2019. https://doi.org/10.3389/fpsyg.2018.01646.

29. Demarin V. The role of arts in enhancement of stroke recovery. Neurosonol Cereb Hemodyn. 2017;13(2):111–4.

30. Torrance EP. Guiding creative talent. Prentice-Hall. 1962. https://doi.org/10.1037/13134-000

31. Torrance EP. Scientific views of creativity and factors affecting its growth. The MIT Press. 1965.

32. Torrance EP. A longitudinal examination of the fourth grade slump in creativity. Gift Child Q. 1968. https://doi.org/10.1177/001698626801200401.

33. Torrance EP. Creativity: just wanting to know. Pretoria, Republic of South Africa: Benedic Books; 1994.

34. Torrance EP. The torrance tests of creative thinking norms-technical manual figural (streamlined) forms A & B. Bensenville, IL: Scholastic Testing Service, Inc.; 1998.

35. Sternberg RJ. Creating a vision of creativity: the first 25 years. Psychology of aesthetics, creativity, and the arts. American Psychological Association; 2006, Vol. S, No. 1. p. 2–12 https://doi.org/10.1037/1931-3896.S.1.2.

36. Sternberg RJ, Lubart TI. An investment theory of creativity and its development. Hum Dev. 1991;34:1–31. https://doi.org/10.1159/000277029.

37. Levitin DJ. This is your brain on music: the science of a human obsession. Plume/Penguin. 2007.

38. Demarin V. Zdrav mozak danas—za sutra. Zagreb: Medicinska naklada; 2017.

39. Sacks O. Musicophilia: tales of music and the brain, revised and expanded edition. Vintage; Revised & Enlarged Edition. 2008.

40. Campbell D. The Mozart effect. William Morrow; 2001.

41. Schlaug G. Musicians and music making as a model for the study of brain plasticity. Prog Brain Res. 2015;217:37–55. https://doi.org/10.1016/bs.pbr.2014.11.020.

42. Raichlen DA, Bharadwaj PK, Fitzhugh MC, et al. Differences in resting state functional connectivity between young adult endurance athletes and 2. Healthy controls. Front Hum Neurosci. 2016;10:610.

43. Dietrich A. The cognitive neuroscience of creativity. Psychon Bull Rev. 2004;11(6):1011–26. https://doi.org/10.3758/BF03196731.

44. Herrmann N. The creative brain. Train Dev J. 1981;35(10):10–6.

45. MacLean P. The triune brain. Science. 1979;204:1066–8. https://doi.org/10.1126/science.377485.

46. Herrmann N. The creative brain, first edition, lake lure. NC: Brain Books; 1988.

47. Guilford JP. Creativity. Am Psychol. 1950;5(9):444–54. https://doi.org/10.1037/h0063487.

48. Kneller GF. Introduction to philosophy of education. New York: Wiley; 1971.

49. Guilford JP. Traits of creativity in creativity and its cultivation. Harper and Row. 1959:142–161.

50. Kim KH. The creativity crisis: the decrease in creative thinking scores on the torrance tests of creative thinking. Creat Res J. 2011;23(4):285–95. https://doi.org/10.1080/10400419.2011.627805.

51. Jorlen A. Link:http://adamjorlen.com/2013/06/10/five-types-of-creative-thinking. In: Martin RL, editor. The design of business: why design thinking is the next competitive advantage. Boston, MA: Harvard Business, 2009. Print.

52. de Bono E. The use of lateral thinking. International Center for Creative Thinking. 1967.

53. Colzato LS, Ozturk A, Hommel B. Meditate to create: the impact of focused-attention and open-monitoring training on convergent and divergent thinking. Front Psychol. 2012;3:116. https://doi.org/10.3389/fpsyg.2012.00116.
54. Manna A, et al. Neural correlates of focused attention and cognitive monitoring in meditation. Brain Res Bull. 2010;82(1–2):46–56. https://doi.org/10.1016/j.brainresbull.2010.03.001.
55. Jaarsveld S et al. Intelligence in creative processes: an EEG study. Intelligence. 2015;49:171–8. https://doi.org/10.1016/j.intell.2015.01.012
56. Dietrich A, Kanso R. A review of EEG, ERP, and neuroimaging studies of creativity and insight. Psychol Bull. 2010;136(5):822–48. https://doi.org/10.1037/a0019749.
57. Schwab D, et al. The time-course of EEG alpha power changes in creative ideation. Front Hum Neurosci. 2014. https://doi.org/10.3389/fnhum.2014.00310.
58. Jauk E, Benedek M, Neubauer AC. Tackling creativity at its roots: evidence for different patterns of EEG alpha activity related to convergent and divergent modes of task processing. Int J Psychophysiol. 2012;84(2):219–25. https://doi.org/10.1016/j.ijpsycho.2012.02.012
59. Fink A, Benedek M. EEG alpha power and creative ideation.Neuroscience & Biobehavioral Reviews. 2014; 44: Pages 111–123 https://doi.org/10.1016/j.neubiorev.2012.12.002.
60. Fink A, Grabner RH, Benedek M, Neubauer AC. Divergent thinking training is related to frontal electroencephalogram alpha synchronization. Eur J Neurosci. 2006;23:2241–6. https://doi.org/10.1111/j.1460-9568.2006.04751.x
61. Fink A, et al. The creative brain: investigation of brain activity during creative problem solving by means of EEG and FMRI. Hum Brain Mapp. 2009;30(3):734–48. https://doi.org/10.1002/hbm.20538
62. Benedek M, Schickel RJ, Jauk E, Fink A, Neubauer AC. Alpha power increases in right parietal cortex reflects focused internal attention. Neuropsychologia. 2014;56(100):393–400. https://doi.org/10.1016/j.neuropsychologia.2014.02.010
63. Dietrich A. How creativity happens in the brain. Palgrave Macmillan. 2015. https://doi.org/10.1057/9781137501806
64. Andreasen NC. The creating brain: the neuroscience of genius. Dana Press. 2005.
65. Sawyer RK. Explaining Creativity. Oxford University Press. 2012.
66. Šarić I. Brainy drawings. Zagreb's Institute for the Culture of Health. 2014.
67. Kandel E. Reductionism in art and brain. Science. 2016. https://doi.org/10.7312/kand17962.
68. Goldberg E. Creativity: the human brain in the age of innovation. 2018.
69. Abraham A. The neuroscience of creativity. Cambridge University Press; 2018.

Artificial Intelligence and Brain Health

Filip Derke

Introduction to Artificial Intelligence

Artificial intelligence (AI) is the field of science concerned with the study and design of intelligent machines. The goal of AI is to build machines that are capable of performing tasks that we define as requiring intelligence, such as reasoning, learning, planning, problem-solving and perception. The field was given its name by computer scientist John McCarty, who along with Arvin Minsky, Nathan Rochester and Claude Shannon, organized the Dartmouth Conference in 1956 (McCarty et al. 1955) [1, 2]. Over the last 60 years, AI has grown into a multidisciplinary field involving computer science, engineering, psychology, philosophy, ethics, medicine and more [3]. For the people unfamiliar with AI, the thought of intelligent machines may at first conjure images of human-like computers or robots, such as those described in SF stories. Because of often misunderstanding in public media and general population about what AI is, and consequential fear of it, we should consider to maybe find another word that will better explain all the benefits of AI such as augmented intelligence.

Intelligent machines have several advantages over human healthcare professionals. Modern expert systems and other intelligent machines can help with highly complex tasks and do so with greater efficiency, accuracy and reliability than humans are able to do. AI technologies can also greatly improve self-care options for persons seeking self-treatment or health-related information [1]. In Europe, nearly 45 million citizens reside in areas without a sufficient number of mental healthcare practitioners to meet the needs of those communities [4]. Now, on the one hand, there are so many AI applications that have been deployed in high-income country contexts, while on the other, the usage of those applications in resource-poor settings remains relatively nascent. Luckily, there are signs that

F. Derke (✉)
Department of Neurology, Clinical Hospital Dubrava,
Av. Gojka Šuška 6, HR-10000 Zagreb, Croatia
e-mail: fderke@me.com

© Springer Nature Switzerland AG 2020
V. Demarin (ed.), *Mind and Brain*, https://doi.org/10.1007/978-3-030-38606-1_2

this is changing. In 2017, the United Nations (UN) convened a global meeting to discuss the development and deployment of AI applications to reduce poverty and deliver a board range of critical public services.

The practical application of AI technologies and techniques in behavioural and mental healthcare is a rapidly advancing area that presents many exciting opportunities and benefits [4]. In the context of health, especially mental and brain health, AI has the potential to become affordable, advantageous and end-user (i.e., patients) friendly toll in mission to improve health, recognize 'red flag' symptoms and involve patients following the concept about 'person centre medicine'.

Neurobiology of Creativity and Machine Learning

For better understanding and building 'the strong AI', computer scientists asked neuroscientist to explain to them about how does the brain work. If they plan to build machines with intellectual ability that is indistinguishable from that of human beings, they need to know brain physiology and fundamentals of neuroscience [1]. Maybe the best example of the brain complexity is to try to explain neurobiology of creativity.

The first point to make is that creativity is not to be found in one distinct section of the brain or a singular clump of nerves behind your left ear. The process is shared across a number of regions and involves a concerto of brain-wide neuronal activity. This makes sense when considering the variety of tasks that exercise our creative bent. Completing a jigsaw or a sudoku involves a certain amount of creative thought, but the sections of the brain relevant to carry out these types of tasks will be different from those involved in designing an art installation or forging the perfect sentence to explain a complex concept. The general consensus is that the creative process has two stages. The first stage is the free flow of experimentation and the creation of a new concept or work of art. The second phase involves rehearsing, editing and assessing the final product as it evolves into the final piece [5].

In this short example, you can see that we don't have one centre in the brain for creativity, there is no one move or one action in 'doing' creativity. The creativity is the process performed in our brain because of activity of many brain regions, many neural connections combined with learned skills, past memories and creating new pathways. The creativity is one of the most complex processes in our brain, evidence-based medicine cannot yet give us answers on questions about the process, but we can follow the process, try to find explanations and learn from it.

The last one, learning by watching how brain works, is the most helpful in machine learning. On the one side, computer scientists try to make artificial data network based on brain networks and pathways. On the other side, neuroscientists try to explain neuroplasticity, clinicians (mostly neurologists and psychiatrists) try to find the way how to encourage the 'brain potential' in recovering or slowing down the progression of diseases. All in all, all those scientists try to answer on the question: How it works? Some of them in that mission start from the beginning of process, some of them from the end try to explain it, some of them create theories about how we cannot give the answers, etc.

Machine learning (ML) is a core branch of AI that aims to give computers the ability to learn without being explicitly programmed. ML has many subfields and applications, including statistical learning methods, neural networks, instance-based learning, genetic algorithms, data mining, image recognition, natural language processing, computational learning theory, inductive logic programming and reinforcement learning [6].

Artificial neural networks (ANN) are a type of ML technique that simulates the structure and function of neuronal networks in the brain. With traditional digital computing, the computational steps are sequential and follow linear modelling techniques. In contrast, modern neural networks use nonlinear statistical data modelling techniques that respond in parallel to the pattern of inputs presented to them. As with biological neurons, connections are made and strengthened with repeated use [6]. We will present just one example in everyday usage of AI based on artificial neural networks. It is the Clinical decision support system or CDSS, which is subtype of expert system that is specifically designed to aid in the process of clinical decision-making. Traditional CDSS rely on preprogramed facts and rules to provide decision options. However, incorporating modern ML and ANN methods allows CDSS to provide recommendations without preprogramed knowledge. Machine learning as a method of collecting the data provide a useful qualitative computational approach for working with uncertainties that can help mental healthcare professionals make more optimal decisions that improve patient outcomes [6].

Virtual Doctor and Therapeutic Games

Virtual doctors (VDs), in the context of AI, and therapeutic video games are highly sophisticated technologies, which represent a culmination of decades of research and development across a number of sub-disciplines of commuter science, medicine, engineering and psychology [7].

Virtual doctor, known in some literature as intelligent virtual agent, is computer-controlled characters that can interact with humans. Implementing believable and engaging virtual agents requires a board range of computational resources and capabilities, including natural language processing (understanding), speech processing (speech understanding), dialog management, fundamental artificial intelligence (automated reasoning, machine learning), affective computing (emotion recognition, expression of emotions, etc.), human–computer interaction and computer graphics (animation and 3D modelling) [7].

Therapeutic games, known as serious games (in contrast with entertainment games), have been developed across diverse contexts and cover a wide range of domains, including language and cultural skills, most academic subjects, sports and motor skills, a variety of corporate and business skills, etc. In healthcare, games are used both for training and education purposes for healthcare professionals, and for therapeutic and educational purposes for patients and clients. Examples are education about diet and exercise, coaching to increase physical activity, pain management, social skills training and more [7]. Today on the market, there are more than 50% of all serious games intended for brain health activities, 'brain-fitness', memory improvement, early recognition signs of mental diseases. One of the examples is *Ricky and the Spider*, game designed to treat OCD in children based on a cognitive-behavioural approach. The players are provided psychoeducation about the condition and are supported in creating a hierarchy of symptoms and provided with opportunities for simulated exposure and response prevention [8]. Another example is Virtual Iraq, the virtual reality therapeutic game. The game implements a VR-mediated exposure therapy approach to the treatment of PTSD in veterans returning from the Iraq and Afghanistan wars. To sum up, therapeutic games have a unique ability to engage the user by providing an immersive environment, directly engaging the user in a pleasurable activity, and by providing immediate positive feedback via a well-defined reward structure [7]. There is much research that needs to be done. In order to optimize intelligent machine systems for interaction with care seekers, research on human–computer and human–robot interaction in the context healthcare is required.

Ethical Issues and Artificial Intelligence Technologies in Mental Health

Will AI provide more fair and objective decisions than humans, who are limited by our own personal experience and biases? Or will they collect and even amplify human prejudices, embedding discrimination within healthcare systems? If the training data isn't representative, or the goals inappropriately chosen, then the resulting AI tool could be deeply inequitable. Any technology that has the potential to track human behaviour and collect personal data must address privacy and data security considerations. These considerations have been codified into the legal system via regulation [9, 10].

Machine learning algorithms being used outside of healthcare have been criticized for discriminating based on race, gender, age and religion, while chatbots have been tricked into propagating hate speech. Artificial intelligence can 'learn' the wrong values and even become self-fulfilling. For example, an algorithm for helping with job hiring decisions might simply reward people who have the same background as those in the historical recruitment data, reinforcing its bias with every decision.

The 'black box' nature of neural networks makes it particularly hard to truly assess whether an AI is biased. Worse still, machine learning is very good at identifying proxies for characteristics, such as predicting race and socioeconomic group from names and postcodes. Tech companies such as IBM, Facebook and Microsoft are all creating tools to help identify bias in algorithms [9, 11, 12].

All things considered, one of the hardest tasks is to balance between ethical issues about AI and ethical issues about mental health. With all those biases, inequalities and unfairnesses, clinicians and scientists should pay attention to those three considerations: (1) clinicians will need to be confident that the decision support tools are valid for the patient in front of them, not just the specific group that made up the training data; (2) algorithms can lead to wrong assumptions based on incomplete data and (3) doctors learn from errors through reflection and changing future practice, regarding to this statement, we can ask ourselves how can we stop algorithms from reinforcing their own behaviour when they make mistakes?

Conclusion

AI holds tremendous promise for transforming the provision of healthcare services in resource-poor settings. Advances in AI help humankind to raise the bar of its potential for creativity and function. Technological advances can help enhance our intellectual and physical capabilities and increase our overall productivity. AI is bringing about a paradigm shift for behavioural and mental healthcare. Mental healthcare professionals, physicians, ethicists, engineers and others must cooperate and work together to accomplish the full potential of what AI and other technologies can bring to behavioural and mental healthcare.

References

1. Howell WC. Engineering psychology (2001). Retrieved from https://www.sciencedirect.com/topics/psychology/engineering-psychology.
2. Filipovic-Grcic L, i Derke F. Artificial intelligence in radiology (2019). Rad Hrvatske akademije znanosti i umjetnosti. Medicinske znanosti, (537=46–47), 55-59. https://doi.org/10.21857/y26kec3o79.
3. Feldman J. Artificial intelligence (2001). Retrieved from https://www.sciencedirect.com/topics/biochemistry-genetics-and-molecular-biology/artificial-intelligence.
4. Academy of royal medical colleges. AI in healthcare (2019). Retrieved from https://www.scribd.com/document/420817825/AI-in-Healthcare.
5. 6.Newman T. The neuroscience of creativity (2016, February 17). Retrieved from https://www.medicalnewstoday.com/articles/306611.php.
6. Bhavsar P. Machine learning in transportation data analytics. Machine Learning - an overview | ScienceDirect Topics, 2017. https://www.sciencedirect.com/topics/psychology/machine-learning.
7. Hudlicka E. Virtual affective agents and therapeutic games. In artificial intelligence in behavioral and mental health care (pp. 81–115) 2016. https://doi.org/10.1016/b978-0-12-420248-1.00004-0.

8. Gunter B. The future of television as an entertainment source. Computer Game - an overview | ScienceDirect Topics, 2010. https://www.sciencedirect.com/topics/psychology/computer-game.
9. MacEwen C. Artificial intelligence in healthcare. Readkong.com, 2019. https://www.readkong.com/page/artificial-intelligence-in-healthcare-2801207.
10. Newell S, Simon HA. The logic theory machine: a complex information processing system. Santa Monica, CA: The Rand Corporation (1956).
11. Parsons S, Mitchell P. The potential of virtual reality in social skills training for people with autistic spectrum disorders. JIDR, 46(Pt 5), 430443 (2002)..
12. Affective computing: theory, methods, and applications by Eva Hudlicka, published by Chapman and Hall (2013).

Neuropsychoanalysis—Gluing the Bits Together Again

Anton Glasnović, Goran Babić and Vida Demarin

Neuropsychoanalysis is an emerging interdisciplinary field of research aimed at applying neuroscientific findings to the psychoanalytic theory and vice versa [1, 2]. It links two major conceptual frameworks, neuroscience and psychoanalysis. Each field deals with a distinct subject, although both of them are viewed as two separate aspects of the same matter. While neuroscience studies brain biology, its functions, and structures that can be objectified, psychoanalysis studies subjective mental processes. However, neuroscience was faced with several questions: Can it, through complex neuronal mechanisms, explain not only workings of the brain, but also workings of the mind? And, how likely is it that reducing the human mind to molecular interactions and neural networks could ever account for complex subjective phenomena, such as sensations, perceptions, feelings, thoughts, and consciousness? It was precisely these questions that prompted the merging of the two disciplines.

The aim of this review is to discuss how neuropsychoanalysis originated and to determine its position in modern medicine, as well as to give basic insights into affective neuroscience—a prime mover of this new and exciting scientific discipline. The timing when the merging of disciplines took place is also of importance because the accumulation of new knowledge arrived at a critical point where the unification of neuroscience and psychoanalysis seemed inevitable. One of the major figures in neuroscience, Antonio Damasio, has recently implied this potentiality. He challenges the mind–body dualism, introduced by a philosophical view of the mind introduced by the seventeenth-century thinkers, mainly Rene Descartes, and argues that emotions guide (or bias) human behavior and that in decision-making individuals use not only cognitive, but emotional processes.

A. Glasnović (✉)
Croatian Institute for Brain Research, Salata 12, HR-10000 Zagreb, Croatia
e-mail: anton.glasnovic@cmj.hr

G. Babić
Croatian Psychoanalytic Society, Zagreb, Croatia

V. Demarin
International Institute for Brain Health, Ul. Grada Vukovara 271, HR-10000 Zagreb, Croatia

© Springer Nature Switzerland AG 2020
V. Demarin (ed.), *Mind and Brain*, https://doi.org/10.1007/978-3-030-38606-1_3

According to Damasio, "Descartes' error" is the dualist separation of mind and body, rationality and emotion. Damasio's theory points out the "decisive function of emotions in navigating the endless stream of life's personal decisions" [3, 4].

The other significant dualistic concept named "dual aspect monism" was articulated by Immanuel Kant and Arthur Schopenhauer. During the history of our understanding of the mind, not only philosophers, but also scientists adopted the dualistic view. Jung is one of the prominent figures who emphasized the dual nature of mental and physical states [5]. Freud, on the contrary, adopted the monistic approach in his medical research, but due to the stage of technological development of his time, continued to develop only the theory of mind, namely psychoanalytical method. In his writings, he speculated that one day, as technology advances, the paradigm will have to shift back to the monistic approach [6]. As neuroscience and psychoanalysis continued to advance separately, there was a challenge of bridging the communication gap between them, mainly due to opposing methodological approaches. This task was not accomplished for most of the century. Driven by advances in research design and methodology and attempts to provide the best practice evidence, neuroscientists and psychologists, similar to experts in other medical fields and disciplines, embraced the evidence-based philosophy. At the same time, psychoanalysis continued to progress in its own setting without implementing research methods used in evidence-based medicine in everyday practice. This inevitably led to criticism from other fields and again sparked the old fight between the disciplines for dominance over matters of the mind. For example, Edwin Garrigues Boring, an experimental neuropsychologist, argued that "psychoanalysis does not have experiments, and has neither the control nor the ability to distinguish between semantic specification and the facts" [7]. Eric Kandel, both neuroscientist and psychiatrist, argued that psychoanalysis was far better at generating ideas then testing them, and therefore did not have the ability to progress in the same way as other areas of research of the mind and medicine in general. He explored the possibilities of biology to refresh the psychoanalytic exploration of the mind [8]. On the other hand, psychoanalyst Marshall Edelson holds that biology is irrelevant to psychoanalysis [9], an opinion he shares with many modern psychoanalysts. Despite all these divergent ideas, Mark Solms and several other neuropsychologists who are also psychoanalysts made remarkable progress toward providing a dialogue framework for the two fields, and through their research, a new discipline, neuropsychoanalysis, came into existence at the beginning of the third millennium [2, 10].

As Freud had foreseen and Kandel asserted that the technology and knowledge had advanced to a point where all prerequisites were met for a dialogue of the two disciplines through a monistic approach. The development of new clinical and laboratory methods such as electroencephalography, evoked potentials, genetic methods and imaging methods made it possible to bring neuroscience and psychoanalysis closer together than ever before. One of the most interesting examples of such cooperation is the explanation of hemispacial neglect of the left side of the body after brain damage. The left-sided hemispacial neglect was traditionally thought to be the consequence of the physical damage of the contralateral hemisphere. However, this view was challenged after a detailed observation of several such patients in a psychoanalytic situation. The results of these observations imply a different explanatory mechanism, namely that the phenomenon might be a consequence of various complex defense mechanisms of the unconscious [11, 12]. Another interesting breakthrough in this field came from the research of the dreaming process. According to findings from studies on this subject matter, the thought process is constantly activated during the awake state, as well as during sleep, and the dreams, more or less structured, can appear during every phase of the sleeping state, not only as we usually thought, during the rapid eye movement phase. The most important dream generator is a sufficiently intense arousal stimulus that initiates the dreaming process by deactivating the motor and premotor frontal cortex. Cortical deactivation releases the mesocortical and mesolimbic "seeking systems," and activates the associative cortex of the occipitotemporoparietal junction to form an illusion, which we call a dream [12]. All of this is in absolute concordance with Freud's idea that the dream protects the subject from waking, and that the thought process is continuous during sleep, but that it is converted into dreaming under various stimuli. The dreaming process is not under control of the superego, which is more or less the case in the awake state [13, 14].

Still, some scientists mistrust the idea of neuropsychoanalysis, claiming that neither of the fields will have much to gain from this conjunction. One of the skeptics is the cognitive neuroscientist Marck Ramus, who argues "that the science of the mind already exists, and that is psychology." Psychology already cooperates with neuroscience, therefore he sees neuropsychoanalysis as "nothing more than an attempt to rehabilitate psychoanalysis by giving it a fashionable prefix and by attributing it the merits of other disciplines." Ramus goes one step further and warns about the perils of psychoanalytic treatment in certain cases. He gives the example of unsupported psychoanalytical approach in autism, claiming that "psychoanalysis rejects international classification of mental disorders in favor of their own idiosyncratic ones." By practicing analytical forms of psychotherapy, whose efficacy is not supported by any empirical evidence, psychoanalysts, in his opinion, delay the diagnosis of autism and subsequent educational intervention [15].

It is useful to give a short review on psychoanalytic models of human mind–work used in everyday psychoanalytic practice. There are two fundamental hypotheses in psychoanalysis, which are the principle of psychic determinism—every action has a reaction, and the proposition that "consciousness is an exceptional, rather than a regular attribute of psychic processes" [16]. The forces that get us into action are called "drives" and there are basically two of them: eros (libido) and thanatos (aggression). Human psyche can be topographically divided into Unconscious, Preconscious and Conscious, and structurally into Id, Ego and Superego [17]. Interaction of all of those components results in human behavior. After Freud's death and until now, there were various psychoanalytic schools, mainly gathered around certain prominent figures such as Melanie Klein ("object relations" theory), Donald Winnicott (concepts of "holding", "good-enough mother", "hate in contratransference", and importance of child's play in the psychic development), Wilfred Bion (concepts of "reverie", "container and contained" and "alfa and beta elements"), Karlheinz Kohut (founder of "self psychology" and in-depth analysis of narcissistic patient) and others, and also theories such as "attachment theory" and "intersubjectivity." It is a matter of style of particular psychoanalyst which he/she uses in psychoanalytical setting with his/her patients, but general assumptions are the same for all. When working with the patient, psychoanalyst uses patient's free associations to analyze his/her unconscious thoughts. Also, there is a great deal of information about unconscious that can be revealed through analysis of dreams (as Freud once said: "Dreams are the royal path to the unconscious"), but today, the most important thing in psychoanalysis as a therapeutic method is thought to be the analytic process and the setting itself, as well as the long-enough experience of a good object. If we presume that a patient is fixed in some stage of his psychic development due to an inability of his/her childhood caregiver to satisfy its emotional needs, then the experience of psychoanalyst being a good-enough "emotional container" and good-enough "caregiver" through years of psychoanalysis may lead to the resumption of stopped mental maturation. This is nicely seen when working with the patient, but also hardly believed without being in the analytical process itself during the psychoanalytical education process.

We should now go through the most recent model of the human mind that neuropsychoanalysis has to offer. In his lecture given to American Psychoanalytic Association, in January 2017, Mark Solms, one of the most prominent neuropsychoanalysts indicated these three core claims: "1. The human infant is not a blank slate; like all other species, we are born with a set of innate needs; 2. The main task of mental development is to learn how to meet these needs in the world, which implies that mental disorder arises from failures to achieve this task; 3. Most of our ways of meeting our needs are executed unconsciously, which requires us to bring them to consciousness once more in order to change them." He continues with the statement that "needs are felt and expressed as emotions" and proposes seven basic emotions: Seeking, Lust, Fear, Rage, Panic, Care, and Play [19]. He also says that "the main task of mental development is to learn how to meet our needs in the world." All these claims are in concordance with Freud's assumptions, and at this very moment are scientifically justifiable [20, 21].

So, after a glimpse into some of the opposing arguments, can we predict the future of neuropsychoanalysis? Or maybe psychoanalysis as such? We think that, in order to see things more clearly, a change in perspective is paramount. The subject of psychoanalysis is not the objective reality of an individual, whose features and attributes can be counted and statistically analyzed like in somatic diseases, where one can measure, for example, blood glucose levels or prove the existence of objective markers of hip arthrosis or carotid artery stenosis. Psychoanalysis deals with the feelings that patients have about themselves and the surrounding world. It also deals with the subjective matter of things, whereas neuroscience deals with the objective matter of things. This notion makes psychoanalysis also an art form, and art cannot be quantified like blood pressure or neurotransmitter concentration in the synaptic cleft [22, 23]. However, the practitioner's skill and technique can be evaluated, as well as the level of inherent meaning of the approach, and eventually its intent to fulfill its purpose, that is, to help the patient.

When trying to account for the complexity of the individual psyche, psychology, as a "science of the mind" [15] due to its reductionist concepts and clinical methodology, faces somewhat similar problems. Indeed the psyche is not equal to cognition or even less to consciousness, but is something much wider. In order to comprehend it, and to help individuals suffering from mental disorders, we have to leave our prejudice behind and move away from the narcissistic phase, but this time armed with new technologies and new perspectives. Although the research setting of neuroscience is the lab bench, and the setting of psychoanalysis is the couch, new interesting insights into our mental functioning arrive daily from both of the disciplines. This prompts us to move from brain to bedside, and vice versa in order to gain a deeper knowledge about ourselves.

Declaration of Conflicts None of the authors declare any conflict of interest. Also, part of the text was taken from our publication Glasnović A, Babić G, Demarin V. Psychoanalysis has its place in modern medicine, and neuropsychoanalysis is here to support it. Croat Med J. 2015 Oct;56(5):503–5, but with permission of Croatian Medical Journal.

References

1. Yovell Y, Solms M, Fotopoulou A. The case for neuropsychoanalysis: why a dialogue with neuroscience is necessary but not sufficient for psychoanalysis. Int J Psychoanal. 2015.
2. Panksepp J, Solms M. What is neuropsychoanalysis? Clinically relevant studies of the minded brain. Trends Cogn Sci. 2012;16:6–8.
3. Damasio A. Mental self: the person within. Nature. 2003;423:227.
4. Damasio A. Feelings of emotion and the self. Ann N Y Acad Sci. 2003;1001:253–61.
5. Atmanspacher H, Fuchs CA. The Pauli-Jung and its impact today. Exeter: Imprint Academic; 2014.
6. Freud S. Beyond the pleasure principle. Leipzig, Wien, Zurich: Internationaler Psychoanalyticsher Verlag; 1921.
7. Boring EG. A history of experimental psychology. New York: Appleton-Century; 1950.
8. Kandel ER. Biology and the future of psychoanalysis: a new intellectual framework for psychiatry revisited. Am J Psychiatry. 1999;156:505–24.
9. Edelson M. Hypothesis and evidence in psychoanalysis. Chicago: University of Chicago Press; 1984.
10. Solms M, Lechevalier B. Neurosciences and psychoanalysis. Int J Psychoanal. 2002;83:233–7.
11. Besharati S, Forkel SJ, Kopelman M, Solms M, Jenkinson PM, Fotopoulou A. The affective modulation of motor awareness in anosognosia for hemiplegia: behavioural and lesion evidence. Cortex. 2014;61:127–40.
12. Turnbull OH, Fotopoulou A, Solms M. Anosognosia as motivated unawareness: the 'defence' hypothesis revisited. Cortex. 2014;61:18–29.
13. Solms M. Neurobiology and the neurological basis of dreaming. Handb Clin Neurol. 2011;98:519–44.
14. Malcolm-Smith S, Solms M, Turnbull O, Tredoux C. Threat in dreams: an adaptation? Conscious Cogn. 2008;17:1281–91.
15. Ramus F. What's the point of neuropsychoanalysis? Br J Psychiatry. 2013;203:170–1.
16. Brenner C. An elementary book of psychoanalysis. New York: Anchor Books; 1955.

17. Freud S. The Ego and the Id. Vienna: Internationaler Psycho-Analytischer Verlag; 1923.
18. Solms M. The scientific standing of psychoanalysis. Br J Psychiatry Intl. In press.
19. Panksepp J. Affective neuroscience. Oxford University Press. 1998.
20. Solms M. The scientific standing of psychoanalysis. BJPsych Int. 2018;15(1):5–8.
21. Paulson S, Hustvedt S, Solms M, Shamadasani S. The deeper self: an expanded view of cosciousness. Ann N Y Acad Sci. 2017.
22. Konopka LM. Where art meets neuroscience: a new horizon of art therapy. Croat Med J. 2014;55:73–4.
23. Braš M, Đorđević V, Janjanin M. Person-centered pain management—science and art. Croat Med J. 2013;54:296–300.

Healing Beyond Mind and Body

Igor Mošič

Healing is defined as restoring health. Further, health is simply defined as the state of physical and mental well-being. Health supports life, and life is the most valuable treasure for every living being. Life is a marvel we receive at birth, as a gift. Among us, humans, each and everyone of us holds our life very dear and extremely important. With regard to cherishing our own human life, every single one of the seven billion people is the same. To understand life better, we can investigate its opposite, and ask the questions: "What is the opposite of life? How specifically it looks when a living being does not have a life?" There is nothing to see, hear, smell, taste, or experience, there are no feelings, no perceptions, no concepts. There is nothing to win, or lose, nothing to build, or create, nothing to destroy, nothing to write about, nothing to be happy, or sad about.

The job of science is finding the truth because we want to make life on our planet better, healthier, safer, and happier for all living beings, which means there is a constant expansion of the scope of collective scientifically verifiable wisdom that benefits all life. We strive to do that in a way better than any other time in human history.

"The truth about the mind, as well as the truth about any other phenomenon has two entities: one entity is a superficial mode of appearance, and other entity is its deep mode of being. These two are called, respectively, conventional and ultimate truth" [1].

Two entities mean that conventional and ultimate truths are of the same nature, which means they are just a different aspect of the same phenomenon. They are different, but inseparable at all times.

From the perspective of the conventional truth, the flower, rose, appears to exist solidly, established independently, out there in the garden. While from the perspective of the ultimate truth, the rose depends on many causes, parts, as well as the name, using which we outerly label the collection of all those conditions.

I. Mošič (✉)
NAM Emotional Hygiene Technology, Podvoljak 8, 51000 Rijeka, Croatia
e-mail: igor@n2sed.hr; igor@yabadooit.com

© Springer Nature Switzerland AG 2020
V. Demarin (ed.), *Mind and Brain*, https://doi.org/10.1007/978-3-030-38606-1_4

The rose exists only nominally as a linguistic label "the rose", which is imputed on the basis of its parts. If the rose would exist the way it appears, there would be an objective rose out there, free from conventions, naming, and labeling. But that kind of objective rose does not exist independently from the label "the rose". It comes to existence as "the rose" because we outerly label it that way, not the other way around. There was never "the rose" out there, because that name came in dependence on our mind.

Who labels the collection of various parts of the rose? The mind. The mind is not a part of the rose, and therefore the rose is said to be outerly labeled. It follows that our experience of reality does not exist independently out there, but it is in the nature of our mind.

If we search for the rose among the parts of the rose, or among many conditions that made rose possible, we will be unable to find the part, or the condition named "the rose".

If we apply this logic to all internal and external phenomena, and search for them, they will remain unfindable because they are labeled only through conventions.

Conventional truth experiences all internal and external phenomena as solid, independent objects, just as they appear to the sense consciousnesses. Ultimate truth experiences all internal and external phenomena as a collection of dependently originated appearances, empty of any form of independent existence. Therefore, phenomena are dependent on various causes, conditions, and parts from the side of the phenomena, and the mind which knows the phenomena and conventionally labels them through thinking and naming.

All phenomena can exist only in two ways: independently or dependently. If the phenomenon exists independently, it cannot exist dependently, and vice versa. Independent and dependent existence of the phenomenon are mutually exclusive.

If the phenomena would exist independently, it would not have parts and would not need an outer label in the form of the name. Further, no causes or conditions could ever influence that phenomenon. On the other hand, phenomenon that exists dependently, has parts, causes, and conditions that influence it, and there is an outer label in the form of the name, which distinguishes it from all the other phenomena.

Hence, if we try to find an external or internal, solidly existent, independent phenomenon, we will not be able to find it.

It follows that there is no internal or external phenomenon which is not dependently originated.

Therefore, our experience of the world is also dependently originated phenomenon. Basically, it depends on an observed object, an empowering condition, and the mind—the observer. For example, for visual consciousness of the rose to arise, three main conditions have to be met: the rose nearby, functioning eye organ, and the consciousness that has the potential to be aware of the rose. It follows that the object, the organ, and the consciousness are intimately interrelated or interconnected.

"In general, the mind can be defined as an entity that has the nature of mere experience, that is, clarity and knowing" [2].

It follows that every consciousness is a knower which is incontrovertible with regard to its main object, by the power of experience. We will focus on two types of consciousnesses: sense and mental consciousness. There are five main types of sense consciousness—eye, ear, nose, tongue, and body, which feed mental consciousness with the information about the specific object of perception. Based on that, six types of mental consciousnesses can arise: manifest form held by the mind, manifest sound held by the mind, manifest taste held by the mind, manifest smell held by the mind, manifest tangible object held by the mind, and manifest phenomenon held by the mind.

Sense consciousnesses know objects by the power of experience and depend on the physical sense power—eye, ear, nose, tongue, and body sense power located within the sense organs [3].

Mental consciousness knows its object by the power of experience and depends on the mental faculty—previous moment of consciousness.

Definitions of sense and mental consciousnesses reveal an extremely important distinction between them, which presents a strong leverage that is needed to uncover the mind's healing potential. For every well-intended health professional, healer, or the person who at the moment experiences illness, this leverage will become clear throughout the text as we go deeper into the logical reasoning about mind, body, and life.

Clear understanding about ultimate and conventional truths and different types of consciousnesses, brings numerous benefits to the process of restoring health.

If health professionals, healers, and people suffering from illness do not apply any critical analysis, they believe illness exists the way it appears—as a solid, independent entity. Falling into this trap brings suffering because the more we do it, the more illness becomes solid. Even though we want to restore our health, the more we believe in conventional appearance of illness, we are transforming merely nominal terms into independently existing, solid thing which creates more pain and suffering.

The more we continue with the process of solidifying illness, we are in fact reinforcing what we apparently wish to eliminate. We are reinforcing illness, and weakening restoring health.

Developing strong reasoning based on the understanding of inseparable nature of conventional and ultimate truths enables one to see illness as dependently arisen phenomenon, depending on many causes, conditions, parts, and the mind which observes, knows, and labels it.

Since our mind is an important condition of dependently arisen phenomenon of illness, convictions which emerge as a result of this reasoning will fundamentally change the way we experience illnesses and the relationship we have with it. If we fixate on solidifying diagnosis, and independent existence of illness, it will drastically limit our possibilities and the outcomes will mostly rely on statistics, completely ignoring the reality and the power of mental consciousness. As opposed to that, our ability to see illness as a dependently arisen phenomenon will tap into the vast space of numerous healing possibilities and favorable outcomes. It is said that once we are able to do it, the progress in mind's healing abilities becomes unlike anything else because it is supported by our sharp intelligence.

Now, to go deeper and take the understanding of that process to a more subtle level, we can critically analyze the reality of mind, body, and human life, the way it exists, not the way it appears to exist.

Since life is our most precious asset, we will first familiarize with our human life. First of all, we know that human life exists, there is no doubt about it, but the basic question is "How specifically does human life exist?"

Does it exist as an independently existent phenomenon, as an object that we can point at? At first, it may appear that life exists independently by its own side, but true scientists cannot be satisfied with appearances. True scientists have developed sharp intelligence which they use to go beyond appearances in their quest for truth.

Human life is just a conventionally used term, because no matter how hard we try to point at human life, we will be unable to do it. Let us try and find what specifically is the human life: Is the blood pressure human life, or is it a blink of an eye, or maybe movement of a hand? Is walking human life, is it talking, dancing, running, smiling, breathing, sexual arousal, heart rate, body temperature? The more we investigate, the more we realize that we cannot point at any specific evidence and say: "This is human life." The only thing we can point at when searching for human life is the basic vital signs (breathing rate, blood pressure, heart rate, and body temperature) and all the other evidence of life. Neither of those signs is specifically the human life because all of those signs make life possible, but none of them is human life. As much as we search for it, we cannot find the human life between the vital signs and other evidence of human life. Further, human life is a singular term while the vital signs and other evidence are plural. If upon search we were unable to find the specific location or the human life, we can conclude that human life is not an independently existent phenomenon because it depends on vital signs, other causes, conditions, and the mind perceiving it. Human life is so rich and full of vital signs, causes, conditions, and other evidence of human life, but at the same time, the human life is empty of itself because none of those vital signs, causes, conditions, and evidence are a human life.

The human life is a name designated to the attribution of vital signs, other evidence, and conditions of life. Thinking and naming are the products of the mind. Vital signs and other evidence of life are experiences of the body. Therefore, we can conclude:

Human life is an experience of a constant interplay between mind and body.

To prove that statement as valid, we will employ the critical analysis for the mind and the body, just as we did for the human life.

Human mind exists, there is no doubt about it. Just the fact that you are reading this, proves its existence. The question is not, does the human mind exist, but "How specifically does the human mind exist?" If we want to find the mind, we might ask: Is eye consciousness the mind? Is it the ear, taste, smell, body, or mental consciousness? Are feelings, or perceptions the mind? Or is it maybe different mental states like anger, jealousy, doubt, kindness, generosity, compassion, or enthusiasm? Are feelings, or perceptions of pleasure and pain the mind? Which one of those is the mind? Just as with human life, the more we search for the mind, the more evidence there is of its existence, but none of them can be pointed and called the mind. Upon search, we were unable to find the mind among the elements of the mind, so we can conclude that the mind is not an independently existent phenomenon because it depends on different types of consciousnesses, feelings, perceptions, or mental states. Let us investigate further by even questioning the way different types of consciousnesses, feelings, perceptions, or mental states exist.

"According to research fellow Petr Janata of the Center for Cognitive Neuroscience at Dartmouth College, music is such a wanted stimulus: it is not particularly important for survival, but still something inside us yearns for it. By means of fMRI, Janata and his collaborators not only discovered and mapped the areas of the brain associated with melody, but they even confirmed that in every listener those mapped areas differed from one music listening session to another, which indicates the variability or dynamics of the topography of the mapped areas. In other words, every time a person hears the same melody, the same nerve bundle perceives it differently" [4].

With regard to what does the consciousness of sound arise? With regard to the sound, which means that even the consciousness of sound is not an objectively, independently existent phenomenon because it depends on the sound as an object of consciousness. "Therefore, without an object of consciousness remaining close by, we can definitely say that there is no consciousness that apprehends it" [5].

The mind is so rich and full of evidence of its existence—different types of consciousness, feelings, perceptions, and mental states, but at the same time, human mind is empty of itself because none of those evidence is the mind. Just as the mind is not independently existent phenomenon, even the parts upon which the mind is designated are not independently existent objects because they also depend upon different types of causes and conditions, and they are all labeled with different names.

"How does the human body exist?" Does it exist just as it appears, as a unit, independent, and objective phenomenon? In our quest for correct knowledge of the phenomenon, we will once again employ the critical analysis of the body. Is our head the body? Are eyes, mouth, nose, hands, fists, shoulders, legs, calves, feet, or knees the body? Is our heart the body? Can the body be found it our organs, muscles, glands, molecules, atoms, or the parts of the atoms? Are the blood and other body liquids human body? Can the body be found within the breath or the body temperature? The more we search, the more evidence there is of a human body, but none of those evidence is the human body. If the human body would be existent in the parts and conditions that make the body, then there would be as many bodies as there are parts and conditions, which is absurd. Upon search we were unable to find the body among the parts and conditions that make the body, so we can conclude that even the body is not an independently existent unit because it depends upon different parts and conditions.

It is apparent that the mind consists only out of non-mind elements, upon which the term mind is designated. The body consists only out of non-body elements, upon which the term body is designated. The life consists only out of non-life elements, upon which the term life is designated. Hence, the reasoning power emerging from this critical analysis becomes the cause of correct intellectual understanding of dependent nature of mind, body, and human life, as opposed to independent existence.

It means that by controlling the causes and conditions involved in the interplay between the mind and the body, we can directly influence and shape the product of that interplay—the quality of human life.

We approached our quest for the truth with critical analysis using logical reasoning, observing mind, body, and human life from the outside, using the third person methodology. Intellectual understanding of the dependent nature of mind, body, and human life emerged from the third person methodology process.

Our study and contemplation about the mind should not become just mere collecting of new information, but we should assimilate this understanding on an emotional level. The best way to do that is through the practice of meditation.

Meditating is a process of exploring the interplay between mind and body. Through meditation, we are familiarizing with clear and knowing nature of mind, and with positive states of mind. Meditation helps us to develop subjective experience, and that is called first person methodology.

Meditation practices can generally be divided into two groups: mental stabilization or mindfulness and analytical contemplative practices. Within those two groups, there are certain specific practices for accessing mind's healing abilities. Detailed analysis of those meditation practices is beyond the scope of this text, and if you want, you should learn it from other books, or even better, directly from the teacher with sharp intelligence and kind heart.

Modern medical science is mostly based on a third person methodology, meaning that we look at the patient and disease from the outside, fixing it based on the different measurements and analysis. Amazing advances in treatment illness, managing chronic diseases, and health prevention have been made due to third person methodology.

Until recently, modern medical science gave very little recognition and awareness to how the mind and the subjective experience of the patient may play a role in the physical health. During the last two centuries, people like Sigmund Freud, Carl Gustav Jung, and Milton H. Erickson were leading the field of exploring the influence of the mind and the patient's subjective experience on mental health as well as human well-being. Taking into account the patient's subjective experience is the basis of the first person methodology.

Those two methodologies; third person methodology which is dominant and first person methodology, developed parallel to each other for many years, not really overlapping with regard to patient treatment.

The last few decades are the beginning of a new and exciting era for all of us because scientists asked the question which is one of the greatest mysteries of human existence: "How specifically does the brain work?" Exploration of that question blurred the boundaries between third and first person methodology of scientific approach. The field where those two methodologies merge is neuroscience, where the tremendous advances have been made with regard to understanding of the human brain through one of the greatest discoveries of our times—brain neuroplasticity [6].

"Neuroplasticity, also known as cortical mapping, challenges the idea that brain functions are fixed in a certain time. It refers to the ability of the human brain to change as a result of one's experience, that the human brain is 'plastic' and 'flexible'" [7].

Using modern technology devices like functional magnetic resonance imaging (fMRI), it is possible to scientifically prove such an amazing discovery as neuroplasticity.

His Holiness the Dalai Lama says: "…evidence is gradually emerging from science, especially psychology and neuroscience, to suggest that it is possible to achieve meaningful change in our emotional and behavioral patterns through conscious effort….recent discovery of what is called 'brain plasticity' may well offer a scientific explanation for this possibility of meaningful change. Researchers have observed that the patterns and structures of the brain can and do change over time in response to our thoughts and experiences. Moreover, scientists are now able to observe the interaction between those parts of the brain associated with higher cognitive activities such as rational thought (in the prefrontal cortex) and those parts known as the limbic system, including the almond-shaped amygdala, which are associated with our most primitive instinctual, and emotional reflexes" [8].

Many scientific research have been made with regard to different types of meditation, especially mindfulness meditation and the benefits of it regarding the improvement of the practitioner's physical and mental health.

Professor of clinical psychology at Oxford University, Willem Kuyken and his colleagues in a Mindfulness-based cognitive therapy for depression, proved that effectiveness of mindfulness-based cognitive therapy (MBCT) in the prevention of depressive relapses is highly relevant for clinical practice and justifies MBCT as a clinically relevant alternative to maintenance antidepressant medication [9].

The results of more than three decades of research have shown the many positive effects that mindfulness can have on health, improving the quality of life both in the general population and in clinical populations, as it is presented by Francesco Pagnini, Assistant Professor at Catholic University of Milan and Deborah Phillips, Research Associate in Psychology at Harvard University in their article Being mindful about mindfulness [10].

Among other benefits, meditation induces equanimity of the mind, a neutral mental state, because we train not to get attached to our conceptual imputations, which means we create space and harmony between cognition and emotions. In our everyday life that is rarely the case. If we imagine mind as a mirror, then sensory perceptions of our everyday life would be appearances on that mirror. Appearances on the mirror are not the basic nature of the mirror; clarity is. Therefore, through certain types of meditation, we want to familiarize ourselves with the clarity of the mind, not the appearances—constructs of the mind. We are familiarizing with the knower, not the known. With the observer, not the observed.

Up to this point, we have clearly developed an understanding of causality, through logical reasoning of dependent existence of mind, body, and human life, because they all arise depending on causes, conditions, and parts. They do not come into being from nothing, or independently.

"Just as material things possess their substantial causes and their contributory conditions, mental phenomena do as well. Our feelings, our thoughts, and emotions, all of which make up our consciousness, have both substantial causes that turn into a particular moment of cognition, and contributory factors that may be physical or mental...a substantial cause must be substantially commensurate with its effect. A physical phenomenon could therefore not serve as the substantial cause of a moment of consciousness, as the nature of clarity and knowing is not physical" [11].

"Thus, each moment of consciousness serves as a substantial cause of our subsequent awareness" [12].

It follows, "the mental consciousness does not rely exclusively upon a specific physical organ support the way the five sense consciousnesses do. The substantial cause of mental consciousness is the previous moment of that consciousness itself. Generally speaking, this arises to some extent on the impressions produced by the physical experience of the senses. So, indirectly we could say that the organ support for the mental consciousness is the momentum of all of the consciousnesses connected with sense experience. But the mental consciousness is that which generates and experiences all of the varieties of emotion and thought that we know—attachment, aversion, bewilderment, apathy, pride, jealousy, feelings of joy and delight, feelings of sadness, feelings of faith and compassion, etc.— all of these different emotional states and all of the thoughts connected with them are varieties of experiences of the sixth or mental consciousness. Now, as these various thoughts and emotions pass through our minds, they transform and influence that consciousness itself. But not only that, they also affect the five sense consciousnesses. For example, when you are very sad and you look at something, you will perceive it as sad, or unpleasant. If you look at the identical object when you are happy, you will see that same thing as pleasant. And if you look at it when you are angry, you will see, again, the same object as entirely different. This is a very simple example of how the mental consciousness is particular and our mind in general affects our experience of sense objects and the sense consciousnesses and the sense organs themselves" [13].

With regard to the mind's innate healing ability, the essential point is that mental consciousness does not get the information about the body directly from the body, but through the body consciousness, which is a sense consciousness. In the same way, mental consciousness does not get the information about the external environment from the environment directly, but through the five sense consciousnesses—eye, ear, smell, taste, and body consciousness. It follows that internal and external phenomena do not communicate with mental consciousness directly, but through the five sense consciousnesses.

It follows, that the way we experience our body is determined by mental consciousness. It is a completely subjective category, dependent on the mind.

Conversely, mental consciousness communicates directly with the body and environment, which means it does not need the five sense consciousnesses to do that.

Studying, contemplating, and meditating about the nature of the mind are means of using human intelligence to train our mental consciousness in developing the correct view of how all internal and external phenomena exist, which can transform our mind into the instrument of healing.

How is that possible? Since it is proved that mindfulness meditations induce equanimous states of mind which positively affect our quality of life, we can take that one step further and ask: "What would happen if we familiarize ourselves with the positive states of mind?" We can define the positive states of mind by thinking about their opposites, negative states of mind—anger, hatred, jealousy, doubt, arrogance, excessive attachment, grasping, ill will, cruelty, dullness, laziness, dishonesty, and the like. The common denominator between them is that they make us and others agitated, sad, and disturbed. Conversely, positive states of mind contribute to inner-peace and create a sense of well-being for oneself and others. The most important positive states of mind are compassion, kindness, generosity, moral conduct, patience, enthusiasm, and altruism. Just as light and darkness cannot coexist, the positive and negative states of mind cannot coexist, also. Positive states of mind are the antidotes to negative states of mind.

Who experiences positive or negative states of mind? Us, but more accurately, it is our mental consciousness. This understanding is extremely important.

Therefore, once the mind is equanimous and calm, we can proceed with mind training or meditation and familiarize ourselves with the qualities like compassion and kindness. Compassion is the positive orientation toward the well-being of others and the sincere wish to help them reduce or eliminate their pain, challenges, or suffering if we can. If we cannot, we should at least refrain from hurting them more.

"While science has made great strides in treating pathologies of the human mind, far less research exists to date on positive qualities of the human mind including compassion, altruism, and empathy. Yet these prosocial traits are innate to us and lie at the very centerpiece of our common humanity. Our capacity to feel compassion has ensured the survival and thriving of our species over millennia. For this reason, the Center for Compassion and Altruism Research and Education (CCARE) at Stanford University School of Medicine was founded in 2008 with the explicit goal of promoting, supporting, and conducting rigorous scientific studies of compassion and altruistic behavior. Founded and directed by Dr. James Doty, Clinical Professor of Neurosurgery, CCARE is established within the Department of Neurosurgery. To date, CCARE has collaborated with a number of prominent neuroscientists, behavioral scientists, geneticists, and biomedical researchers to closely examine the physiological and psychological correlates of compassion and altruism. CCARE investigates methods for cultivating compassion and promoting altruism within individuals and society through rigorous research, scientific collaborations, and academic conferences" [14].

Our human life does not exist as an independent phenomenon, which means it is dependent upon vital signs, causes, and conditions. None of those vital signs, causes, or conditions are "our human life". Meaning, our human life consists only out of "non-our human life" elements upon which the term "our human life" is designated. What are the constituent elements of the set "non-our human life"? As we have already mentioned those are our vital signs and other sensory measurable evidence, but also different causes like food that we eat, the air that we breathe, water, sunlight, etc. Since we constantly associate with other people that greatly influence our well-being, "other's human lives" are also the elements of the set "non-our human life", isn't it? Further, "lives of all other living beings" are also the elements of the set "non-our human life".

There is no phenomenon that is not dependently arisen. Therefore there is no phenomenon that is not empty [15] (of independent exisstence).

It follows, if we want to make our human life better, healthier, safer, and happier, the only way to do that is to use the interplay of our mind and body, and positively influence the constituent elements of "our human life". One of the most important constituent elements of "our human life" is "other's human life", or on a higher logical level "all life".

This kind of logical reasoning shows we are completely safe when we support the marvel of our human life by helping other living beings support their life, with compassion, kindness, enthusiasm, and wisdom as opposed to not helping them. If we do help, we are completely safe. The life will prevail. All life without exception.

Therefore, the essential ingredients of this wisdom are kindness and compassion. Here, we are talking about the wisdom that realizes how things exist in reality, as opposed to how things appear to exist.

Gradual development of that wisdom is the prerequisite that serves as the cause for uncovering and activating innate healing potential of the mind. The progress in developing that wisdom depends on the study of the science of the mind, contemplating about it using reason and logic, and then integrating newfound convictions on an emotional level, through the practice of meditation.

Health and life are conventional terms which complement and support each other. Therefore, whatever is true for life, could be applied to health and healing.

Illness is not independently established, we can say that it is empty of independent existence. That does not mean, the illness does not exist, on the contrary, it exists, but just as a collection of dependent arisings coming from the side of the body, as well as from the side of the mind, or the observer.

Without realizing deeper mode of ultimate reality, we are trapped in the labyrinth of independent, solidly existent appearances; such as illness. Mind that is trained with study of the science of mind, contemplation, and meditation will realize incontrovertible truth with regard to the true mode of existence of illness. The truth is based on reality, not appearances. The benefit of that process, unlike anything else, weakens the solidity of illness, and opens vast space for healing ability of the mind to emerge in its full potential.

For the health professionals, this reasoning is a profound way to help patients to restore health, by seeing them the way they really exist, not the way they appear to exist as "ill human beings". For people who experience illness, this reasoning can serve as an inspiration and sound logical base to unsolidify illness and start training the mind to heal the body.

To present all in a condensed way, "consciousness pervades the experience of its objects and the experience of the organs, if the consciousness is transformed, or one's mode of experience of consciousness is transformed into the pure realization of final mode of abiding or the ultimate truth, then the appearances of the objects, and also the organs themselves, will become pure and sacred" [13].

Pure and ill are mutually exclusive, meaning those two modes of being cannot coexist. In this way, developing strong, logically supported conviction based on the same nature of ultimate and conventional truth, and assimilating it on an emotional level through the specific practices of meditation, could benefit not only mind, but also the body.

Familiarizing with subjective realization of the ultimate mode of abiding of illness will weaken the power of illness to bring suffering, because our experience of illness comes into existence in dependence on our mental consciousness. Therefore, the experience of illness also dissolves through the power of the wisdom that realizes reality as it exists, as opposed to the way reality appears to exist. That wisdom emerges from the mental consciousness through the study of science of the mind, contemplation, and meditation. Hence, mental consciousness, which has the ability to develop that kind of wisdom, pervades the experience of our body and our mind. Depending on that kind of reasoning, we can infer that innate healing potential of the mind is always accessible and healing is possible.

Since healing is a hidden phenomenon, we used logical reasoning to prove that the pure view, based on the indivisibility of conventional and ultimate truth, is valid and supported by reality.

In conclusion, despite all possible logical fallacies and inconsistencies which stem from my lack of higher knowledge or special wisdom, this paper is my humble effort to single-handedly approach human ability to heal mind and body, using mind's innate healing potential. It is my wish that thoughts presented here serve as a spur for the rest of the scientific community with more resources and wisdom to keep expanding the field of research, so we could gain more practical understanding about the ways to uncover and access the mind's healing potential, for the benefit of all.

References

1. Lama Dalai. The Buddhism of Tibet and the key to the middle way. Boston: Snow Lion; 2002. n.p.
2. Lama Dalai. What is the mind. In: Mind Science, Goleman D, Thurman RF, editors. Somerville, MA: Wisdom Publications; 1991. Reprinted with the permission in the Mandala Magazine; 1995. https://www.lamayeshe.com/article/what-mind.
3. Rinbochey Lati. Mind in Tibetan Buddhism. Ithaca, NY: Snow Lion Publications; 1986. p. 33.
4. Demarin V, Bosnar Puretić M. The brain and art. In: Croatia and Europe: Medicine, Science & Arts—Scientific and Professional Papers, Zagreb: Croatian Committee of the European Association of Arts Medicine; 2007. p. 49.
5. Shantideva. A guide to the Bodhisattva's way of life. Dharamsala: Library of Tibetan Works and Archives; 1979. p. 175.
6. Jinpa T. Mental illness recovery through mindfulness meditation. DVD, Dharamsala: Men-Tsee-Khang; n.d.
7. Zavoreo I, Basic-Kes V, Demarin V. Stroke and neuroplasticity. In: Periodicum biologorum, vol. 114, no. 3. Zagreb: Hrvatsko prirodoslovno društvo; 2012. p. 395.
8. Lama Dalai. Beyond religion. India: HarperCollins Publishers; 2012. p. 114.
9. Kuyken W, Hayes R, Barrett B, et al. Mindfulness-based cognitive therapy for depression. The Lancet. 2016;387. http://www.thelancet.com/journals/lancet/article/PIIS0140-6736(16)00660-7/.
10. Pagnini F, Philips D. Being mindful about mindfulness. Lancet. 2015;2. http://www.thelancet.com/journals/lanpsy/article/PIIS2215-0366(15)00041-3/.
11. Lama Dalai. A profound mind. London: Hodder & Stoughton; 2011. p. 70.
12. Lama Dalai. An open heart—practicing compassion in everyday life. New York: Little, Brown and Company; 2001. n.p.
13. Khenchen Thrangu Rinpoche. Medicine Buddha teachings. Ithaca, NY: Snow Lion Publications; 2004. p. 6.
14. The center for compassion and altruism research and education [Internet]. Stanford, CA: Stanford University School of Medicine, Mission & Vision, n.d. http://ccare.stanford.edu/about/mission-vision/.
15. Nagarjuna. Mulamadhyamakakarika—Chapters 26, 18, 23, 24, 22, and 1 [Internet]. In: Translator and editor Geshema Kelsang Wangmo. Chapter 24, Verse 19. http://tushita.info/wp-content/uploads/2018/07/Nagarjunas-Mulamadhyamakakarika-Translation-by-Geshe-Kelsang-Wangmo-2018.pdf.

Primary Headache from a Psychosocial Perspective

Bojana Žvan and Marjan Zaletel

A headache is a perception of pain in the head. In primary headaches such as migraine and tension headaches, we do not find any source for nociceptive activity such as a haemorrhage or inflammation. We believe that biopsychosocial factors are involved in headache perception, acting interactively through the circle of biological, psychological and social items, which are inseparable. The final consequence is distress. From the biological view, it represents the organism's response to stress such as environmental conditions, and its adaptation to changes. However, we are now living in a more complex social environment and thus we need to adapt to the challenges of social situations. Transaction models deal with psychobiological responses to stress that arise from the disproportion in perceived requirements, as well as the individual's personal and social capabilities to meet the requirements. Distress most commonly occurs when perceived requirements exceed the perceived capabilities or can also occur when the perceived sources exceed requirements. Therefore, the distress from a headache creates a more subjective assessment in an individual as to how they perceive the requirements that define the distress level, rather than an objective assessment of the requirements and abilities [1]. Nevertheless, distress is not always due to physical stress, but social situations in which they cannot cope, can also be a source of social stress. We also need to raise awareness about the social aspects of the pain in society itself, as well as to reduce negative social processes that create 'bad' social situations.

Interoceptive Model of a Headache

Interoception is a sense of the internal state or homeostasis of the body and it includes visceral as well as nociception [2]. Physical stress on the body produces a disturbance of the internal environment and to the physiological responses needed

B. Žvan (✉) · M. Zaletel
Clinical Department of Vascular Neurology, University Medical Centre Ljubljana,
Zaloska 2, 1000 Ljubljana, Slovenia
e-mail: bojana.zvan@kclj.si

to maintain the body's homeostasis. The nervous system is important with regards to coordinating the physiological response. On the other hand, the brain can predict future physiological states from past experiences. The representation of the body and the homeostatic states constitutes the basis for our predictions. To manage our life in accordance with external demands without overstraining, the predictions should be aligned with our sensations. The mismatch between predictions and sensations produces the prediction error which means a 'surprise' for the nervous system, which therefore could evoke a behaviour or autonomic response to minimise the prediction error [3], and, to correct the prediction error or stress on the body, we consume additional energy. Distress may arise when the perceived capabilities do not meet the perceived requirements and thus, it is a subjective state. Personality features, especially catastrophising, may promote distress and headaches.

The predictions on future states could be wrong but very precise. For example, the precise prediction of pain under certain circumstances situations can evoke a headache despite nonspecific sensory input. The result is a headache, which reflects distress because of the unpredictable situation and we feel unpleasant. In this state, the active and perceptual inference [4] may operate with autonomic response such as trigeminovascular vasodilatation of cerebral arteries with calcitonin gene-related peptide (CGRP) in order to eliminate the prediction error. However, the autonomic response could reinforce the perception of the headache. In addition, the sickness behaviour develops to diminish the sensory inputs, and as the episode progresses the headache should gradually subside. The source of wrong predictions could be the social environment, cultural features, individual experiences from the childhood period or examples in the family. Mood disorders may highlight the predictions and feelings of pain.

From the neurophysiological point, the headache represents the nociceptive activation of trigeminocervical complex in pain perception [5]. This model does not distinguish between sensation and perception. Other models include the sensitisation of second and higher order neurons due to excessive stimulation, which leads to hyper sensibility states. The others prefer to regard the neuromodulatory pain system in the brain stem as the origin of the headache. Nevertheless, all of the models should explain the influences of psychosocial factors on patients with headaches.

Social Stress

Social stress is the most common type of stress experienced by people on a daily basis and affects us more than other types of stress [6]. By nature, people are social beings and have a fundamental need and desire to maintain positive social connections. Social connections offer a caring environment that supports individuals in an emotional and physical sense and promote a feeling of social inclusion. They lead to the creative success and performance of an individual. All actions that destroy or threaten social connections with other people cause social distress in an individual [7]. Social stress arises from an individual's relationships or connections with other people and from the social environment in general. According to the concept of emotion assessment, distress occurs when an individual assesses a situation as being personally important and perceives that they are no longer able to solve a specific situation. Researchers have identified social stress and stressors in various ways. Social stressors were defined as a set of characteristics, situations, episodes, and behaviours related to psychological and physical strains and tensions, and are of a social nature [8]. Wadman [9] defined it as a feeling of discomfort and fear that individuals experience in a social situation and is related to the behaviour or tendency to avoid potentially stressful social situations. This is in accordance with the interoceptive model in which the perception of pain and emotions arise as a consequence of homeostatic disturbance and cognitive factors. Ilfield [10] defined social stressors as situations in everyday social roles that are generally problematic and unwanted.

Among the social stressors, we distinguish three types: life events that are defined as sudden, severe life changes that require rapid adaptation, such as a sudden injury or sexual assault, chronic stress is defined as continuous events that require long-term adaptation such as divorce or loss of a job; and daily stress as small events that require adaptation throughout the day.

Individual's Response to Stress

Each of us is different and unique. Therefore, our characteristics as an individual, such as personality, desires, motivation, knowledge and emotions, importantly affect our perceptions and social behaviour. At the same time, our behaviour significantly affects the social situation and people with whom we have daily contact. We are living in a form of equilibrium with our social environment. Social stressors create turbulence in such balance and produce social stress that causes distress for some individuals. This could evoke our perceptions of distress and pain, and induce behaviours to minimise these effects, and even to change our social environment. Our social influence happens through the thought processes with which other people change their thoughts, feelings and experiences. Psychologist Lewin defined the group's influence as the individual variable and the situational variable that are known as interaction with the individual [11]. Evolutionary adaptation has provided us with two basic motivations [12]. The first one is related to us (ego). Freud divided personality into three layers, named "id"—unconscious, "ego"—conscious, and "superego", which contains a consciousness and a morality acquired in society. "Id" consists of two needs: a need for sexuality (Eros) and destruction (Tanatos). The libido pushes a human to fulfil these two needs. Ego (me) is everything we are aware of. Motivation is related to us (ego) protects us, increases our ego and the ego of people who are psychologically close or similar to us. The second is the motivation that is related to a social situation, namely, connecting with others, accepting, and being accepted. These two motivations are related to the care of oneself and taking care of others. The conflict between different parts of a personality, usually ego and superego, can lead to psychological stress and distress, which creates headaches.

Distress as a Trigger Factor of Primary Headaches and as a Risk for Chronic Headaches

The retrospective studies almost entirely show that distress is the most common trigger for migraine and tension-type headaches in adults. Various researches confirm distress as a trigger for primary headaches, especially migraines [13]. Recently published meta-analysis of the existing studies on connections among distress and primary headaches, clearly showed that the most important triggering events for primary headaches are distress and sleeping disorders [14]. Studies have reported important connections between distress and headaches, but they also emphasised individual differences in relation stress headaches. Recent major research examined distress levels and the number of days with headaches per month, in 3-month periods over a period of 2 years. Distress intensity was related to headaches in individuals who had tension-type headaches, migraines, and accompanying migraines. In tension-type headaches, there was a stronger relationship between distress and headaches in younger participants [15]. Another piece of research studied the decrease of distress as a possible triggering factor for migraines, which in the literature is described as a weekend headache. The research establishes that the distress level is not related to headaches, but a decrease of distress from one evening to the next is related to a higher incidence of migraines [16]. One of the studies actually reported on the protective role of relaxation after distress in migraines. Researches that deal with distress as a setting for headaches, include analyses of life events, everyday problems, and distress in relation to headaches. Researches that studied a relationship between stressful life events and headaches used various measurements and obtained different results [17]. When using an inventory of life events and a questionnaire about life experiences, it found small, but important differences, which indicated a higher number of stressful life events in persons suffering from headaches. In studies related to everyday life problems, they uniformly report that persons with headaches experience more everyday problems than control groups [18]. Researches into distress consistently report on higher stress in persons with headaches than in control groups. Holm presented that individuals with tension-type headache assess distress events more negatively than the control groups [19]. When the potential influence of a stressful event was ambiguous, people with headaches perceived this event as being more significant and that they had less control over it. They found that persons with headaches had fewer effective strategies for coping with stressful events, since they relied on relatively ineffective stress management strategies, such as avoidance and self-blaming, and used less support from their surroundings than the control group [20]. Also, Hassinger reports that persons suffering from migraines are more likely to respond to stress with unreal desires, self-criticism, isolation, and negative thinking in comparison to the control group. Numerous researches revealed that people with headaches have less support from family and their surroundings compared to control groups [21]. Martin and Theunissen reported about the differences among persons suffering from headaches and control groups in terms of social support [18]. Martin and Soon discovered that persons with headaches are significantly less satisfied with the support available to them, and they received lower assessments in all four types of functional support (assessment, self-esteem, belonging, and tangibility). Findings show that clinics and researchers should pay more attention to the social dimension of headaches [22]. Eskin and co-workers reported that patients suffering from tension-type headaches and migraines experience a higher level of stress and a greater lack of help from their surroundings, in solving problems, compared to control groups [21]. Santos discovered that migraines are related to the influence of negative events on life and lower support levels from the surroundings. Financial problems and hospitalisation revealed both direct and independent relationships with diagnosis and migraine frequency. Also, the death of a relative was independently and strongly related to migraines [23]. Many authors assumed that stress was a risk factor for chronic headaches, but evidence for this assumption is limited. D'Amico reports that in 44.8% of the cases in their sample, a stressful event was related to the transition of episodic headaches into chronic headaches. Minor events such as everyday problems have a bigger role in transforming headache than major events. The authors interpret this in a way that patients with transformed headaches react to stress differently, and not that these individuals have been more exposed to major stressful life events [24]. Considering pre-emptive prior headache factors, for many years they claimed that persons suffering from headaches have a certain personality profile. They described them as tense, obsessive perfectionists with a non-flexible personality and holding grudges, which they cannot express nor resolve. In an overview containing findings from more than a hundred researches, Blanchard and co-workers concluded that the data did not point to headaches being related to a certain personality type, but persons with headaches had more frequently experienced psychological distress [25].

Cognitive Factors

When managing patient with primary headaches it is important to evaluate cognitive factors which encompass thoughts, beliefs, expectations, and individual's principles. We are using these when we connect with our social environment. The influence of cognition is obvious when patients occupy themselves with behaviours to decrease the probability of a migraine episode. It includes taking medications and using coping strategies for migraine attacks. Most important are locus of control and self-efficiency. The locus of control is the degree to which an individual perceives that an event is under his or her personal control [26]. The two extremes are complete external and internal locus of control (LOC). The first meaning that the control of headaches is completely accomplished by external factors such as tablets or doctors, and the second is when someone believes the event is under his or her control. It is typical that patients with frequent headache have more external then internal LOC. That's why we should always ask the patients what, or who, is the most important thing in controlling their headaches. They were shown that internal LOC was related to a better treatment outcome and less disability [27]. Low internal LOC is related to behaviours to reduce headache frequency such as avoiding headache triggers or taking medication to eliminate headaches leading to an intense headache. Higher levels of depression were detected in patients with high external LOC who believe that they cannot control their headache by accidental factors. For optimal outcomes, the patients should have high internal and low external LOC.

Self-efficiency (SE) relates to patients' beliefs that they can successfully act in the process of preventing or alleviating their headaches. It depends on individuals' past experiences. High SE is associated with less anxious emotions [28] and increases autonomic arousal. Both concepts, LOC and SE are linked to the prediction of headaches in certain situations and responses to stress in the interoception headache model.

Negative Emotions and Precision of Predictions

Affects associated with pain and headaches encompass negative effects (NA) and emotional states such as depression, anxiety and anger. They can trigger the headache or have an influence on the intensity of the pain. NA is inseparably connected with pain because they are aversive. NA can influence the probability for the oncome, and experiencing, of headaches, their intensity and the headache-related disabilities [29]. In terms of the predictive coding concept, NA has an influence on the precision of our predictions. NA comprises a basic unpleasant defensive motivation system that exacerbates a headache.

Anxiety is an emotional state of excessive worry, fear, or apprehension. It is associated with negative thoughts of being unable to control the situation and the headache. It is a form of distress with emotional feelings of incompetency and a feeling of powerlessness when patients exposed to a stressor become stressed, which then heightens their anxiety, and thus creates a vicious anxiety/stress cycle. This is anticipated anxiety. Anxious feelings are one of the most common headache triggers and individuals with headaches are more anxious than persons without pain [30].

Another emotional state connection to chronic pain and headaches is depression. It is a clinical syndrome described by feelings of sadness, despair, emptiness, and loss of will. A more common form is a dysphoric feeling, which has symptoms like depression but is transitory. It was found that a dysphoric feeling is more common in patients with headaches compared to those without headaches [31]. According to the interoception model the dysphoria increases the precision of predictions, and the prediction error as to why the stress is higher and the likelihood of experiencing distress and headaches is higher. In addition, hopelessness, a thought associated by low internal LOC and low SE and heightened pessimism, often accompanies dysphoric feelings and increased headache-related disabilities [32].

Anger is an effect associated with displeasure and it is perceived in the situation when an individual's well-being is endangered. From the sociological point it could arise when the well-being of others is endangered, especially those with whom we identify [33]. The anger expression can be divided into anger in and anger out expressions. Anger out expressions are socially unacceptable. It is known that patients with headaches are more prone to hold their anger in than those without headaches [34]. The patients who hold their anger have increased pain intensity and a failure to express anger leads to more disability [35]. Overall anger in or out has negative influences on communication and cause more social stress.

Biopsychosocial Approach to Headache Treatment

According to the interoceptive model of headaches, in order to decrease headache frequency, intensity and disability we should change the high precision predictions which produce excessive, and sometimes unneeded, stress, and consequently distress. Social stress, due to wrong predictions, is known as a trigger for headaches. Therefore, clinics should pay more attention to the social aspects of headaches and include them in the treatment process. Recognising and managing psychosocial factors could be crucial for the same patients. Educating the patient about how social triggers influence a headache attack, and how managing triggers can reduce the number of headache attacks, increases internal LOC. Keeping regular diaries of potential headache triggers allows the patient to see how triggers are related to headaches [36]. Teaching skills for self-managing triggers could increase SE. Individuals with catastrophic fears about being exposed to a potential trigger will need to become aware that triggers can set the stage for headaches, but exposure to a trigger does not mean they will experience a debilitating headache. It is necessary to do this in combination with taking medications for headache treatment. Reinforcing the adherence to medication could increase SE and the perception of headache control, which can decrease NA related to headaches. To manage social stress in some cases, more decisive behavioural forms, such as more assertive behaviour, are useful. We should pay attention to the exclusion processes in patients' working environments and in social environments in general. Exclusion is recognised as a serious threat and motivates individual to a defensive posture. It can produce excessive social stress. Nowadays mobbing in the workplace is a common form of social stress. Overall, inadequately managed NA increases the individual's risk of experiencing more headache attacks, more intense headache pain, and more headache-related disabilities. Early recognition and management of the emotional state are mandatory in primary headache management.

Conclusions

Pain and also headaches are not just a consequence of the sensations of biological processes but also consequences of cognition and emotional factors. Social stress due to social interactions is the most frequent form of stress. Social situations together with the personality features of a headache sufferer can produce social stress, which can lead to distress, headache perception and NA. Chronic headaches are importantly related to psychosocial factors. Therefore, in analysis of the causative factors and risk factors, it is important to consider social stress and recognise mechanisms that trigger and maintain headaches. We must be aware that social processes themselves can lead to emotional states, such as depression and anxiety, as well as sleeping disorders that form a vicious circle together with headaches. This can deteriorate the social status and the individual's role in society. Therefore, sometimes, a social worker should be included in the treatment of a patient with chronic headaches.

References

1. Bao S. Mechanical stress. Handb Clin Neurol. 2015;131:367–96.
2. Craig AD. How do you feel? Interoception: the sense of the physiological condition of the body. Nat Rev Neurosci. 2002;3:655–66.
3. Edwards MJ, Adams RA, Brown H, Pareés I, Friston KJ. A Bayesian account of 'hysteria'. Brain. 2012;135:3495–512.
4. Seth AK. Interoceptive inference, emotion, and the embodied self. Trends Cogn Sci. 2013;17:565–73.
5. Goadsby PJ, Holland PR, Martins-Oliveira M, Hoffmann J, Schankin C, Akerman S. Pathophysiology of migraine: a disorder of sensory processing. Physiol Rev. 2017;97:553–622.
6. Prenderville JA, Kennedy PJ, Dinan TG, Cryan JF. Adding fuel to the fire: the impact of stress on the ageing brain. Trends Neurosci. 2015;38:13–25.
7. Shiban Y, Diemer J, Brandl S, Zack R, Mühlberger A, Wüst S. Trier social stress test in vivo and in virtual reality: dissociation of response domains. Int J Psychophysiol. 2016;110:47–55.
8. McEwen BS, Bowles NP, Gray JD, Hill MN, Hunter RG, Karatsoreos IN, Nasca C. Mechanisms of stress in the brain. Nat Neurosci. 2015;18:1353–63.
9. Wadman R, Durkin K, Conti-Ramsden G. Social stress in young people with specific language impairment. J Adolesc. 2011;34:421–31.
10. Ilfeld FW Jr. Current social stressors and symptoms of depression. Am J Psychiatry. 1977;134:161–6.
11. Heims S. Kurt Lewin and social change. J Hist Behav Sci. 1978;14:238–41.
12. Solms M. A neuropsychoanalytical approach to the hard problem of consciousness. J Integr Neurosci. 2014;13:173–85.

13. Lipton RB, Pavlovic JM, Haut SR, Grosberg BM, Bu D. Methodological issues in studying trigger factors and premonitory features of migraine. Headache. 2014;54:1661–9.
14. Yokoyama M, Yokoyama T, Funazu K, Yameshita T, Kondo S, Hosoai H. Association between headache and stress, alcohol drinking, exercise, sleep, and comorbid health conditions in a Japanese population. J Headache Pain. 2009;10:177–85.
15. Schramm SH, Moebus S, Lehmann N, Galli U, Obermann M, Bock E, Yoon MS, Diener HC, Katsarava Z. The association between stress and headache: a longitudinal population-based study. Cephalalgia. 2015;35:853–63.
16. Lipton RB, Buse DC, Hall CB, Tennen H, De Freitas TA, Borkowski TM, Grosberg BM, Haut SR. Reduction in perceived stress as a migraine trigger: testing the "let-down" hypothesis. Neurology. 2014;82:1395–401.
17. Martin PR. Stress and primary headache: review of the Research and Clinical Management. Curr Pain Headache Rep. 2016;20:45.
18. Martin PR, Theunissen C. The role of life event stress, coping and social support in chronic headaches. Headache. 1993;33:301–6.
19. Holm JE, Lamberty K, McSherry WC 2nd, Davis PA. The stress response in headache sufferers: physiological and psychological reactivity. Headache. 1997;37:221–7.
20. Hassinger HJ, Semenchuk EM, O'Brien WH. Appraisal and coping responses to pain and stress in migraine headache sufferers. J Behav Med. 1999;22:327–40.
21. Eskin M, Akyol A, Çelik EY, Gültekin BK. Social problem-solving, perceived stress. Scand J Psychol. 2013;54:337–43.
22. Martin PR, Soon K. The relationship between perceived stress, social support and chronic headaches. Headache. 1993;33:307–14.
23. Santos IS, Brunoni AR, Goulart AC, Griep RH, Lotufo PA, Benseñor IM. Negative life events and migraine: a cross-sectional analysis of the Brazilian Longitudinal Study of Adult Health (ELSA-Brasil) baseline data. BMC Public Health. 2014;14:678.
24. D'Amico D, Libro G, Prudenzano MP, Peccarisi C, Guazzelli M, Relja G. Stress and chronic headache. J Headache Pain. 2000;1:S49–52.
25. Blanchard EB, Andrasik F, Arena JG. Personality and chronic headache. Progress Experiment Personal Res. 1984;13:303–60.
26. Nicholson RA, Houle TT, Rhudy JL. Norton PJ.com Psychological risk factors in headache. Headache. 2007;47:413–26.
27. Scharff L, Turk DC, Marcus DA. The relationship of locus of control and psychosocial-behavioural responses in chronic headache. Headache. 1995;35:527–33.
28. Martin NJ, Holroyd KA, Rokicki LA. Headache self-efficacy scale: adaptation to recurrent headaches. Headache. 1993;33:244–8.
29. Perozzo P, Savi L, Castelli L, Valfrè W, Lo Giudice R, Gentile S, Rainero I, Pinessi L. Anger and emotional distress in patients with migraine and tension type headache. J Headache Pain. 2005;6:392–9.
30. Marcus D. Headache and other types of chronic pain. Headache. 2003;43:49–53.
31. Nicholson RA, Gramling SE, Ong JC, Buenevar L. Differences in anger expression between individuals with and without headache after controlling for depression and anxiety. Headache. 2003;43:651–63.
32. Gesztelyi G, Bereczki D. Determinants of disability in everyday activities differ in primary and cervicogenic headaches and in low back pain. Psychiatr Clin Neurosci. 2006;60:271–6.
33. Smedslund J. How shall the concept of anger be defined? Theory Psychol. 1993;3:5–33.
34. Venable VL, Carlson CR, Wilson J. The role of anger and depression in recurrent headache. Headache. 2001;41:21–30.
35. Wade JB, Price DD, Hamer RM, Schwartz SM, Hart RP. An emotional component analysis of chronic pain. Pain. 1990;40:303–10.
36. Houle T, Remble T, Houle T. The examination of headache activity using time-series research designs. Headache. 2005;45:438–44.

From Mild Encephalitis Hypothesis to Autoimmune Psychosis

Karl Bechter

At the Beginning an Appreciation of Personalities and Atmosphere of Pula Congresses

The interdisciplinary character of Pula neuropsychiatric congresses involved and encouraged early on the just slowly emerging psycho-neuroimmunological research field, especially its rare clinical psychiatric parts and persons. Myself was first invited to present studies on the potential role of Bornavirus in the etiology of neurological and psychiatric disorders by Gerd Huber: He was the renown disciple of Kurt Schneider, appreciating and continuing Schneider's psychopathology and developing it in some important aspect further to the Basic Symptoms concept [1], intending to better assess and define the biological basis of major psychoses, especially of schizophrenia. The Basic Symptoms were thought to represent symptoms most closely (basically) related to assumed underlying organic causes of major psychoses, whatever these causes might be, e.g., viruses. The respective questionnaires developed by him, Süllwold, Gross, Schüttler and others, tried to systematically and sensitively recognize especially the subjective "basic" experiences of the patient, because these subjective experiences were thought to be lastly and specifically of "organic" origin. Beyond it was assumed, that Basic Symptom might represent especially sensitive symptoms of organic brain disturbances, because the individual might be most sensitive for such himself yet fro minor "organic" dysfunction. Nevertheless, these subjective symptoms might additionally and regularly be shaped by normal psychological individual functioning (as far as normal psychological functions remained conserved despite the organic dysfunction) and beyond be characterized by previous ("non-organic") experiences. The Basic Symptom concept has up to now been further worked out and used by disciples and

K. Bechter (✉)
Department Psychiatry and Psychotherapy 2, University of Ulm, Ulm, Germany
e-mail: karl.bechter@bkh-guenzburg.de

© Springer Nature Switzerland AG 2020
V. Demarin (ed.), *Mind and Brain*, https://doi.org/10.1007/978-3-030-38606-1_6

followers internationally and gained special reputation mainly in schizophrenia studies on disease course and prevention [2, 3].

The personal discussions during the Pula congresses with these outstanding psychiatric researchers and many international renown neurologists and experts from neighboring disciplines were of great value for my own thinking. The Pula faculty was not only open-minded but also critical and rewarding in the best sense. Such atmosphere was very helpful for consolidation of my personal development in the difficult etiology focused research I had chosen at these times, when mainstream opinion in psychiatry was mostly discouraging for my own views.

Only from recent research, there is good evidence that infections frequently may trigger autoimmunity. Another surprising knowledge now is that autoimmunity itself may have rather diverse consequences, including pathogenic and protective roles. Such complex scenario was in the nineties of the twentieth century far from clear or considered highly speculative. Now the psychoneuroimmunology field came into mainstream psychiatric research, which includes my own views, as documented by a new diagnostic entity, introduced most recently after a long process of international consensus finding: Autoimmune Psychosis (AP) was proposed as a term yet in 2017 [4], then picked up under extended perspective and definition by others [5], and in 2019 for the first time established with internationally consented diagnostic criteria and treatment recommendations [6]. AP as defined now appears synonymous to autoimmune mild encephalitis. The Mild Encephalitis (ME) hypothesis was proposed some years ago [7, 8] and repeatedly discussed and updated at Pula congresses [9–11]. The Pula audience was supportive yet in times of broad rejection of ME hypothesis in the field. Thus, it appears justified and fair to outline the story from early ME hypothesis to AP description, on the one hand from a personal perspective including the role of Pula congresses yet mentioned above, on the other hand from an actual researchers perspective on many open questions left in context with ME hypothesis and AP description, with some ideas to inform research to come.

From Mild Encephalitis Hypothesis to First Consensus Description of Autoimmune Psychosis

There is emerging evidence that a subgroup of severe mental disorders (SMI), especially of the broadly defined affective and schizophrenic spectrum, may be ultimately caused by underlying neuroinflammatory and/or autoimmune or immune-pathological processes, outlined in recent reviews [4, 5, 9, 12, 13]. Such scenario was predicted with the Mild Encephalitis (ME) hypothesis [7, 9, 14]: ME was assumed to be elicited by various etiologies including immunopathology/autoimmunity, infections, toxicity (exogenous, endogenous), and trauma, these factors representing safely known risk factors in such scenario. ME was also assumed to possibly link to yet unknown causes; with regard to the latter rather recently discovered mechanisms like autoinflammation, inflammasome activation, and genetic-driven vasculopathy might be considered.

Most important in the early times of ME hypothesis was the admittedly rare observation (only few cases treated by this procedure), that even severe therapy-resistant SMI cases often rapidly improved or fully remitted in few days when treated with CSF filtration, a comparably aggressive immune-modulating therapy used in neurology for the typically infection-induced autoimmune disorder Guillain–Barre syndrome [11, 13–16].

The idea of ME to prevail in a subgroup of SMI but remaining undiagnosed due to insensitive methods [7, 9] was strongly supported from the development in psychiatry after the discovery of NMDAR autoantibodies in neurological patients and the consequent description of Autoimmune Encephalitis (AE) by Josep Dalmau and colleagues in 2007/8 [17–19], a now well established new neurological diagnosis [18]. Actual consensus diagnosis of AE includes a guide to major differential diagnosis and treatment also of psychiatric syndromes [19]. Interestingly, the number of cases diagnosed as (neurological) encephalitis increased considerably with the discovery of a further emerging number of CNS autoantibodies being discovered [17, 20, 21 and others not reviewed here].

This development in neurology reinforced a development in psychiatry to detect CNS autoantibodies in SMI cases, but only in recent years comparably aggressive immune-modulatory treatments (high dose cortisone, intravenous immune globulins, plasmapheresis, rituximab, and others) were used in therapy-resistant SMI cases under the idea of an underlying autoimmune pathogenesis and appear to bring about similar often rapid and striking therapeutic success alike which were seen with CSF filtration (for reviews and single case presentations see [5, 8, 12, 20, 22–25, 29]).

AE presents initially typically with varying psychiatric syndromes and only later the onset of severe neurological symptoms [17–19]. Part of AE cases was known for long and was previously diagnosed as limbic encephalitis (LE), cases being observed mainly in neurological clinics. However, rare cases of LE were for long described also in psychiatry, an aspect especially addressed by Gerd Huber and colleague in catatonic schizophrenias with lethal outcome, where mild neuroinflammation was suggested in very rare and debated post-mortem evidence (reviewed in [7]). But with the discovery of NMDAR autoantibodies, and an emerging number of further previously unknown CNS autoantibodies [20], the size of the neurological AE subgroup increased considerably since, now including also clinical syndromes not previously attributed to encephalitis, e.g., some cases of epilepsy [21, others].

The newly defined neurological AE cases in principle and often fulfill established criteria of encephalitis, but it should be recognized that with the discovery of more and more CNS autoantibodies the defining criteria of AE were extending in parallel and seem still broadening beyond the borders and criteria of previous encephalitis definitions. It may be recognized at this point that the number of the new AE cases would match the ME criteria as originally proposed, which rejected by many for long.

Emerging Clinical and Therapeutic Relevance
of the ME/AP Concept

Interestingly and accordingly, due to still improved laboratory diagnostic approach
to detect CNS antibodies in parallel the spectrum of neuropsychiatric syndromes
not fulfilling previous criteria of neurological encephalitis or the initial description
of AE is emerging since AE cases including now also classical neurological disease cases like epilepsy, and, as predicted with ME hypothesis, also cases of SMI
of various syndromal presentation [5, 6, 21]. So, not only some of the newly diagnosed AE cases would fulfill diagnostic criteria of ME, but also number of cases
presenting with pure psychiatric syndromes without neurological hard signs, a
subgroup of those now newly defined as AP [6]. ME as compared to AE is harder
to diagnose, which is not surprising as being of a purely psychiatric presentation,
thus requiring very careful multimodal evaluation of symptoms and findings (see
detailed case descriptions and reviews in Research Topic Frontiers [8];). Given
this intriguing research field and clinical approach appears to be in an early phase
of knowledge, it may be justified for specialized departments/specialists in the
field to challenge the established psychiatric diagnosis even with yet borderline
diagnostic and treatment approaches, as an emerging number of published cases
demonstrates by dramatic improvements with immune-modulatory treatments
[23, 25–27]. Thus, it appears also justified to further questioning present diagnostic borders in context with the ME hypothesis more generally, because more and
more single cases, when carefully diagnosed with multimodal clinical approach
(including CSF examination and neuroimaging with various MRI and PET methods, EEG, brain biopsy) were successfully treated under AP/ME diagnosis, before
being left to therapy resistance.

ME was proposed to be mainly characterized by severe psychiatric syndromes
often accompanied by neurological soft but not hard signs (see Table in [7]), ME
expected to be diagnosed with improved and emerging diagnostic methods including neuroimaging, laboratory methods, and especially CSF examination. The size
of the hypothetic ME subgroup within SMI cohort remained unclear though with
respect to a mosaic of findings was held possible to have considerable size [9].
Recent rare post-mortem and clinical studies presented evidence of a potentially
considerable size of ME subgroup in schizophrenia, i.e. of about 20% of cases
(compare 28) or up to 60% (compare 29). Such findings could match with clinical CSF findings: Mild neuroinflammation appears from an increasing number
of studies, performed according to defined modern criteria of CSF examination
and interpretation in neurology, involved in a SMI subgroup of 10–15% of cases.
However, in considerably larger subgroup of affective and schizophrenic spectrum therapy-resistant cases some other minor CSF pathologies were found and
confirmed in large cohorts (albumin and protein increase; neopterin increase and
some cytokines, or CSF cell activation), but the exact clinical meaning of these
abnormal CSF findings has still to be better explained/understood [22, 30–33].

Beyond schizophrenia and psychosis a variety of clinical syndromes are still
to be considered in the perspective of ME hypothesis, although may be rare, as

the general unspecificity of etiologies suggests [7, 9]. Such view is supported by emerging findings, for example in rare case of adult depression with chronic fatigue syndrome [34], or in cases in childhood/adolescent psychiatry: An example similar to AP/ME seems to represent the defined PANS/PANDAS syndrome, in which after decennia of research now complicated diagnostic and treatment recommendations including antibiotic and/or immune-modulatory treatment options have been worked out [35, 36]. The categorical classification of PANDAS/PANS resembles AP, thus may be considered a childhood/adolescence subtype of autoimmune ME in etiological or pathogenetic perspective. But herein an interdisciplinary discussion might be required or be useful including such aspects like the use of diagnostic measures also.

ME Hypothesis, AP Diagnosis, and Beyond

ME hypothesis and diagnosis appeared early on rather speculative to many colleagues, nevertheless, it was based on animal models of Bornavirus infection and ME was first diagnosed in Bornavirus seropositive patients presenting in parallel some CSF abnormalities evaluated compatible with ME, to select cases suitable for CSF filtration [11, 13–15]. CSF filtration was before successfully used in neurology for cases suffering from the typically infection-induced autoimmune disorder Guillain–Barre syndrome [16]. Indeed, majority of therapy-resistant psychiatric cases selected by these above criteria for potential application of CSF filtration improved with CSF filtration. The recent consensus definition of AP [6] match widely with the previous ME criteria, but were refined and safer by detection of CNS autoantibodies, which can be assessed now. Despite this important progress from the discovery of CNS autoantibodies, there is still rather good plausibility, that also other ME/AP subgroups may prevail and remain undiagnosed with available antibody assays, respectively that other subgroup may exist which does not produce CNS autoantibodies instead may be characterized by cellular or innate immune pathology, as supported for example by rarely performed brain biopsy. Thus, the tentative AP subgroup might eventually embrace at least three autoimmune respectively immune-pathological subgroups [5]: 1. AP presenting with CNS autoantibodies, 2. AP accompanying systemic autoimmune diseases, 3. AP presenting without (known) CNS autoantibodies.

Other open questions in this research field include the interaction between genes and environment, a cardinal feature of ME hypothesis (compare 7, 9), but which is difficult to enlighten because of the variability and complexity of such interaction. Genes may for example shape the clinical appearance (syndrome) and outcome in case of infection or may predispose to infection-induced autoimmunity or the genetic base may infer low resistance to infection (in latter case may end up in infectious ME or in infectious classical encephalitis). Also, an important role of the inflammatory response system by MHC 4 related genes was recently found for schizophrenia [37], but interpretation of these findings could be various and could also well fit with ME scenario. Other scenarios can be discussed, but such undertaken goes beyond the focused review here.

Conclusion

From early days of ME hypothesis with broad rejection of the hypothesis, there is now surprisingly rapid change to acceptance. An ever broadening application of immune-modulatory treatments in SMI, mainly in cases of the affective and schizophrenic spectrum including bipolar and/or schizoaffective and rarely other syndromes, clearly supports ME hypothesis, though indirect, but in principle as predicted. Nevertheless, ME diagnosis required ultimately the proof of mild neuroinflammation, assumed to prevail especially in active disease stages. But such proof was hampered by the limited sensitivity of available clinical methods and low accessibility of the brain in clinical reality and due to apparent ethical aspects. In addition, mild neuroinflammation required a consensus definition, clinically differentiating mild neuroinflammation from classical neuroinflammation and also from parainflammation (but which may have some overlap) or from stress-induced parainflammation [38]. Such is difficult to achieve, as also demonstrated by the rather limited consensus on established terms in clinical use (compare 38). Even for long-traded clinical terms with recent international consensus criteria established like encephalitis and encephalopathy [24] the limitations of clinical diagnostic precision are apparent [38]. Nevertheless, such consensus criteria, though only case sensitive not theoretically sound, are very important for clinical approach, representing the best available evidence for sound diagnosis and appropriate treatment. Such criteria are also prerequisite for research when applying new potentially more successful treatments, to provide the most responsible approach to patients possible when nevertheless recognizing mayn remaining circumstantial uncertainties. The recent consensus criteria of AP thus represent a landmark achievement for psychiatry, AP as defined now representing a CNS antibody-positive autoimmune subtype of ME. The prevalence of still more ME subtypes remains to being predicted and to become discovered.

References

1. Huber G. Das Konzept substratnaher Basissymptome und seine Bedeutung für Theorie und Therapie schizophrener Erkrankungen. Nervenarzt. 1983;54:23–32.
2. Klosterkötter J, Hellmich M, Schultze-Lutter F. Is the diagnosis of schizophrenic illness possible in the initial prodromal phase to the first psychotic manifestation? Fortschr Neurol Psychiatr. 2000;68(Suppl 1):S13–21.
3. Schultze-Lutter F, Klosterkötter J, Ruhrmann S. Improving the clinical prediction of psychosis by combining ultra-high-risk criteria and cognitive basic symptoms. Schizophr Res. 2014;154(1–3):100–6. https://doi.org/10.1016/j.schres.2014.02.010. Epub 2014 Mar 7.
4. Ellul P, Groc L, Tamouza R, Leboyer M. The clinical challenge of autoimmune psychosis: learning from anti-NMDA receptor autoantibodies. Front Psychiatry. 2017;8:54. https://doi.org/10.3389/fpsyt.2017.00054. eCollection 2017.
5. Najjar S, Steiner J, Najjar A, Bechter K. A clinical approach to new-onset psychosis associated with immune dysregulation: the concept of autoimmune psychosis. J Neuroinflammation. 2018;15(1):40. https://doi.org/10.1186/s12974-018-1067-y.

6. Pollak TA, Lennox B, Müller S, Benros ME, Prüss H, Tebartz van Elst L, Klein H, Steiner J, Frodl T, Bogerts B, Tian L, Groc L, Hasan A, Baune BT, Endres D, Haroon E, Yolken R, Benedetti F, Halaris A, Meyer J, Stassen H, Leboyer M, Fuchs D, Otto M, Brown DA, Vincent A, Najjar S, Bechter K. An international consensus on an approach to the diagnosis and management of psychosis of suspected autoimmune origin: the concept of autoimmune psychosis. Lancet Psychiatry. 2019. In Press.
7. Bechter K. Mild encephalitis underlying psychiatric disorder—a reconsideration and hypothesis exemplified on Borna disease. Neurol Psychiatry Brain Res. 2001;9:55–70.
8. Bechter K, Brown D, Najjar S. Editorial: Recent advances in psychiatry from psycho-neuro immunology research: autoimmune encephalitis, autoimmune encephalopathy, and mild encephalitis. Front Psychiatry. 2019;10:169. https://doi.org/10.3389/fpsyt.2019.00169. eCollection 2019.
9. Bechter K. Updating the mild encephalitis hypothesis of schizophrenia. Prog Neuropsychopharmacol Biol Psychiatry. 2013;5(42):71–91. https://doi.org/10.1016/j.pnpbp.2012.06.019. Epub 2012 Jul 3.
10. Bechter K. CSF diagnostics in psychiatry—present status—future developments. Neurol Psychiatry Brain Res. 2016;22:69–74.
11. Bechter K, Herzog S, Schreiner V, Wollinsky KH, Schüttler R. Cerebrospinal fluid filtration in Borna-disease-virus-encephalitis related schizophrenia: a new therapeutic perspective in psychiatry? Neurol Psychiatry Brain Res. 1998;6:85–6.
12. Najjar S, Pearlman D, Zagzag D, Golfinos J, Devinsky O. Glutamic acid decarboxylase autoantibody syndrome presenting as schizophrenia. Neurologist. 2012;18(2):88–91. https://doi.org/10.1097/NRL.0b013e318247b87d.
13. Bechter K, Schreiner V, Herzog S, Breitinger N, Wollinsky KH, Brinkmeier H, Aulkemeyer P, Weber F, Schüttler R. [Cerebrospinal fluid filtration as experimental therapy in therapy refractory psychoses in Borna disease virus seropositive patients. Therapeutic effects, findings]. Psychiatr Prax. 2003;30 Suppl 2:S216–20.
14. Bechter K. The mild encephalitis-hypothesis–new findings and studies. Psychiatr Prax. 2004;31(Suppl 1):S41–3.
15. Bechter K, Herzog S, Schreiner V, Brinkmeier H, Aulkemeyer P, Weber F, Wollinsky KH, Schüttler R. Borna disease virus-related therapy-resistant depression improved after cerebrospinal fluid filtration. J Psychiatr Res. 2000;34(6):393–6.
16. Wollinsky KH, Hülser PJ, Brinkmeier H, Aulkemeyer P, Bössenecker W, Huber-Hartmann KH, Rohrbach P, Schreiber H, Weber F, Kron M, Büchele G, Mehrkens HH, Ludolph AC, Rüdel R. CSF filtration is an effective treatment of Guillain-Barré syndrome: a randomized clinical trial. Neurology. 2001;57(5):774–80.
17. Al-Diwani AAJ, Pollak TA, Irani SR, Lennox BR. Psychosis: an autoimmune disease? Immunology. 2017;152(3):388–401. https://doi.org/10.1111/imm.12795. Epub 2017 Aug 3.
18. Dalmau J, Graus F. Antibody-mediated encephalitis. N Engl J Med. 2018;378(9):840–51. https://doi.org/10.1056/NEJMra1708712.
19. Graus F, Titulaer MJ, Balu R, Benseler S, Bien CG, Cellucci T, Cortese I, Dale RC, Gelfand JM, Geschwind M, Glaser CA, Honnorat J, Höftberger R, Iizuka T, Irani SR, Lancaster E, Leypoldt F, Prüss H, Rae-Grant A, Reindl M, Rosenfeld MR, Rostásy K, Saiz A, Venkatesan A, Vincent A, Wandinger KP, Waters P, Dalmau J. A clinical approach to diagnosis of autoimmune encephalitis. Lancet Neurol. 2016;15(4):391–404. https://doi.org/10.1016/S1474-4422(15)00401-9. Epub 2016 Feb 20.
20. Bechter K, Deisenhammer F. Psychiatric syndromes other than dementia. Handb Clin Neurol. 2017;146:285–96. https://doi.org/10.1016/B978-0-12-804279-3.00017-4.
21. Geis C, Planagumà J, Carreño M, Graus F, Dalmau J. Autoimmune seizures and epilepsy. J Clin Invest. 2019;129(3):926–40. https://doi.org/10.1172/JCI125178. Epub 2019 Feb 4.
22. Endres D, Perlov E, Baumgartner A, Hottenrott T, Dersch R, Stich O, Tebartz van Elst L. Immunological findings in psychotic syndromes: a tertiary care hospital's CSF sample of 180 patients. Front Hum Neurosci. 2015;9:476. https://doi.org/10.3389/fnhum.2015.00476. eCollection 2015.

23. Endres D, Vry MS, Dykierek P, Riering AN, Lüngen E, Stich O, Dersch R, Venhoff N, Erny D, Mader I, Meyer PT, Tebartz van Elst L. Plasmapheresis responsive rapid onset dementia with predominantly frontal dysfunction in the context of Hashimoto's encephalopathy. Front Psychiatry. 2017;8:212. https://doi.org/10.3389/fpsyt.2017.00212. eCollection 2017.

24. Venkatesan A, Tunkel AR, Bloch KC, et al. Case definitions, diagnostic algorithms, and priorities in encephalitis: consensus statement of the international encephalitis consortium. Clin Infect Dis. 2013;57(8):1114–28. https://doi.org/10.1093/cid/cit458. Epub 2013 Jul 15.

25. Mack A, Pfeiffer C, Schneider EM, Bechter K. Schizophrenia or atypical lupus erythematosus with predominant psychiatric manifestations over 25 years: case analysis and review. Front Psychiatry. 2017;8:131. https://doi.org/10.3389/fpsyt.2017.00131. eCollection 2017.

26. Najjar S, Pahlajani S, De Sanctis V, Stern JNH, Najjar A, Chong D. Neurovascular unit dysfunction and blood-brain barrier hyperpermeability contribute to schizophrenia neurobiology: a theoretical integration of clinical and experimental evidence. Front Psychiatry. 2017;8:83. https://doi.org/10.3389/fpsyt.2017.00083. eCollection 2017.

27. Endres D, Perlov E, Stich O, Rauer S, Maier S, Waldkircher Z, Lange T, Mader I, Meyer PT, van Elst LT. Hypoglutamatergic state is associated with reduced cerebral glucose metabolism in anti-NMDA receptor encephalitis: a case report. BMC Psychiatry. 2015;1(15):186. https://doi.org/10.1186/s12888-015-0552-4.

28. Bogerts B, Winopal D, Schwarz S, Schlaaff K, Steiner J. Evidence of neuroinflammation in subgroups of schizophrenia and mood disorder patients: a semiquantitative postmortem study of CD3 and CD20 immunoreactive lymphocytes in several brain regions. Neurol Psychiatry Brain Res. 2017;23:2–9.

29. Zhang Y, Catts VS, Sheedy D, McCrossin T, Kril JJ, Shannon Weickert C. Cortical grey matter volume reduction in people with schizophrenia is associated with neuro-inflammation. Transl Psychiatry. 2016;6(12):e982. https://doi.org/10.1038/tp.2016.238.

30. Bechter K, Reiber H, Herzog S, Fuchs D, Tumani H, Maxeiner HG. Cerebrospinal fluid analysis in affective and schizophrenic spectrum disorders: identification of subgroups with immune responses and blood-CSF barrier dysfunction. J Psychiatr Res. 2010;44(5):321–30. https://doi.org/10.1016/j.jpsychires.2009.08.008. Epub 2009 Sep 30.

31. Kuehne LK, Reiber H, Bechter K, Hagberg L, Fuchs D. Cerebrospinal fluid neopterin is brain-derived and not associated with blood-CSF barrier dysfunction in non-inflammatory affective and schizophrenic spectrum disorders. J Psychiatr Res. 2013;47(10):1417–22. https://doi.org/10.1016/j.jpsychires.2013.05.027. Epub 2013 Jun 19.

32. Maxeiner HG, Rojewski MT, Schmitt A, Tumani H, Bechter K, Schmitt M. Flow cytometric analysis of T cell subsets in paired samples of cerebrospinal fluid and peripheral blood from patients with neurological and psychiatric disorders. Brain Behav Immun. 2009;23(1):134–42. https://doi.org/10.1016/j.bbi.2008.08.003. Epub 2008 Aug 20.

33. Maxeiner HG, Schneider EM, Kurfiss ST, Brettschneider J, Tumani H, Bechter K. Cerebrospinal fluid and serum cytokine profiling to detect immune control of infectious and inflammatory neurological and psychiatric diseases. Cytokine. 2014;69(1):62–7. https://doi.org/10.1016/j.cyto.2014.05.008. Epub 2014 Jun 7.

34. Bechter K, Bindl A, Horn M, Schreiner V. Therapy-resistant depression with fatigue. A case of presumed streptococcal-associated autoimmune disorder. Nervenarzt. 2007;78(3):338, 340–1.

35. Thienemann M, Murphy T, Leckman J, Shaw R, Williams K, Kapphahn C, Frankovich J, Geller D, Bernstein G, Chang K, Elia J, Swedo S. Clinical management of pediatric acute-onset neuropsychiatric syndrome: Part I-psychiatric and behavioral interventions. J Child Adolesc Psychopharmacol. 2017;27(7):566–73. https://doi.org/10.1089/cap.2016.0145. Epub 2017 Jul 19.

36. Swedo SE, Frankovich J, Murphy TK. Overview of treatment of pediatric acute-onset neuropsychiatric syndrome. J Child Adolesc Psychopharmacol. 2017;27(7):562–5. https://doi.org/10.1089/cap.2017.0042. Epub 2017 Jul 19.

37. Sekar A, Bialas AR, de Rivera H, Davis A, Hammond TR, Kamitaki N, Tooley K, Presumey J, Baum M, Van Doren V, Genovese G, Rose SA, Handsaker RE; Schizophrenia Working Group of the Psychiatric Genomics Consortium, Daly MJ, Carroll MC, Stevens B, McCarroll SA. Schizophrenia risk from complex variation of complement component 4. Nature. 2016;530(7589):177–83. https://doi.org/10.1038/nature16549. Epub 2016 Jan 27.
38. Bechter K. Encephalitis, mild encephalitis, neuroprogression, or encephalopathy-not merely a question of terminology. Front Psychiatry. 2019;9:782. https://doi.org/10.3389/fpsyt.2018.00782. eCollection 2018.

Psychiatric Disorders in Neurological Diseases

Osman Sinanović

Introduction

Psychiatric disorders (PDs) in neurology are more frequent then it verified in routine exam, not only in the less developed but also in large and very developed neurological departments [1, 2]. Furthermore, psychiatric symptoms (PSs) in neurological diseases (NDs) among primary health care physicians and other specialties are often neglected. Anxiety and depression are most common, but hallucinations, delusions, obsessive-compulsive disorder, delirium or confusional state and cognitive disturbances are also frequent comorbidity in many neurological conditions such as stroke, epilepsy, multiple sclerosis (MS), Parkinson's disease (PD), Huntington's disease, and Wilson's disease [3, 4].

Clinical neurologists and psychiatrists have long recognized the frequent occurrence of psychiatric conditions in the context of neurologic (brain) disease. Indeed, this frequent co-occurrence of psychiatric with neurologic symptoms should come as no surprise, since psychiatric disorders, such as schizophrenia and the mood disorders, can be induced by structural brain disease. Presumably, brain dysfunction from conditions that cause neurologic symptoms—such as seizures, and impairments in movement, sensation, speech, or language—also affects areas of the brain that regulate mood, emotion, cognition, and perception [5].

O. Sinanović (✉)
School of Medicine University of Tuzla, Univerzitetska 1, 75000 Tuzla, Bosnia and Herzegovina
e-mail: osman.sinanovic1@gmail.com

School of Science and Technology, Sarajevo Medical School University Sarajevo, Hrasnička cesta 3a, 71210 Sarajevo, Bosnia and Herzegovina

© Springer Nature Switzerland AG 2020
V. Demarin (ed.), *Mind and Brain*, https://doi.org/10.1007/978-3-030-38606-1_7

Stroke

Stroke can be defined as a sudden attack of a specific neurological deficit, caused by thrombosis or hemorrhage in the cerebral circulation, and it is the third leading cause of death in developed countries [5, 6]. Psychiatric syndromes associated with stroke lead to significant psychological distress, functional impairments, poor rehabilitation outcomes, and excess mortality [5, 6]. The most common psychiatric disturbances seen after stroke include cognitive impairment and dementia, depression, mania, anxiety disorders, and pathological laughing and crying—now referred to as involuntary emotional expression disorder [8].

Cognitive deficits of several types have been reported, typically in relationship to the location of brain injury. Left-hemisphere strokes frequently cause dysphasia, whereas right-hemisphere strokes are associated with anosognosia, inattention, impaired spatial reasoning, and neglect syndromes. Motivation, memory, judgment, and impulse control may be affected after frontal stroke. Additionally, brain vascular disease is associated with the emergence of dementia. This can be the result of one stroke affecting a single critical area, such as the thalamus, several strokes affecting areas important to cognition, or chronic vascular insufficiency leading to white matter changes with associated cognitive problems ("vascular cognitive impairment") [9–12].

Stroke is a leading cause of disability. Research and interventions have historically focused on physical disabilities, while cognitive impairment—an important aspect for stroke survivors—has been rather neglected. Even minor stroke affects daily functioning, executive functions, and cognition, consequently affecting participation, quality of life, and return to work. Stroke survivors are at increased risk of developing cognitive impairment. Obviously, the acute tissue damage may affect cognition [12].

A variety of classifications, diagnostic criteria, and descriptive syndromes are used to define post-stroke cognitive impairment (PSCI), but a widely accepted and harmonized terminology is still missing. Post-stroke neuropsychological syndromes overlap—PSCI is responsible for a substantial number of vascular cognitive impairment (VCI) syndromes; PSCI includes the subgroups of PSD and PSCI not fulfilling criteria for dementia.

Stroke is recognized as one of the major causes of adult disability globally. VaD including PSD is the second most common cause of cognitive decline, with only Alzheimer's disease (AD) being more prevalent. The lifetime risk of developing either stroke or dementia at the age of 65 is one in three in men and one in two in women. With changing population demographics, increased life expectancy and improved survival from stroke, the absolute numbers of patients with PSD will increase. However, due to its relationship with stroke incidence, PSD might be reduced with improved stroke prevention [12, 13].

Cerebral Autosomal Dominant Arteriopathy with Subcortical Infarcts and Leukoencephalopathy (CADASIL)

Cerebral small vessel disease (SVD) is an important cause of stroke, cognitive impairment, and mood disorders in the elderly. Besides the common sporadic forms, mostly related to age and hypertension, a minority of SVD has a monogenic cause, among which the most common and best known is cerebral autosomal dominant arteriopathy with subcortical infarcts and leukoencephalopathy (CADASIL). CADASIL provides a unique model for the study of the most prevalent forms of sporadic SVD. CADASIL is caused by mutations in the NOTCH3 gene, which maps to the short arm of chromosome 19 and encodes the NOTCH3 receptor protein, predominantly expressed in adults by vascular smooth muscle cells and pericytes [1]. Thousands of families with CADASIL have now been diagnosed worldwide in many different ethnic groups. The disorder is often overlooked and misdiagnosed. Its minimum prevalence has been estimated between 2 and 5 in 100,000 but may vary between populations [14, 15].

The main clinical features include migraine usually with aura presenting in early adulthood, recurrent subcortical ischemic events, mood disturbances, progressive cognitive impairment mostly affecting executive function, and ac Subcortical ischemic events. Transient ischemic attacks and stroke are reported in approximately 85% of symptomatic individuals and are related to cerebral small vessel pathology. Mean age at onset of ischemic episodes is approximately 45–50 years, but the range at onset is broad (third to eighth decade). Ischemic episodes typically present as a classical lacunar syndrome (pure motor stroke, ataxic hemiparesis/dysarthria-clumsy hand syndrome, pure sensory stroke, sensorimotor stroke), but other lacunar syndromes (brain stem or hemispheric) are also observed. The total lacunar lesion load, symptomatic and asymptomatic, is strongly associated with the development of severe disability with gait disturbance, urinary incontinence, pseudobulbar palsy, and cognitive impairment. Strokes involving the territory of a large artery have occasionally been reported but whether these are coincidental or related to the CADASIL pathology itself is uncertain [15].

Encephalopathy

An acute encephalopathy has been described in 10% of CADASIL patients, and in the majority of these it was the first major symptom, with a mean age of onset of 42 years. It is frequently misdiagnosed as encephalitis, particularly if it is the initial presentation in a patient without known CADASIL. It usually evolves from a migraine attack, including confusion, reduced consciousness, seizures, and cortical signs, with spontaneous resolution. Cognitive impairment in CADASIL first involves information processing speed and executive functions, with relative preservation of episodic memory, and is associated with apathy and depression. The cognitive deficits were initially attributed to subcortical origin. More recently, the cerebral cortex was shown also to be affected in CADASIL by direct mechanisms (cortical microinfarcts), or through secondary degeneration. Memory impairment in CADASIL is probably related to different causes, due to both white matter infarcts with disruption of either cortico-cortical or subcortical networks mainly in the frontal lobe and primary damage of the cortex. In a recent longitudinal study, processing speed slowing was related to a decrease of sulcal depth, but not to global brain atrophy or cortical thinning, suggesting that early cognitive changes may be more specifically related to sulcal morphology than to other anatomical changes. Moreover, studies in mouse models of CADASIL have detected dysregulation of hippocampal neurogenesis, a process essential for the integration of new spatial memory records.

Psychiatric Disturbances

Although the clinical expression of the disease is mainly neurological, CADASIL is also characterized by psychiatric disturbances (20–41%). Apathy and major depression are commonly observed in CADASIL. Also, bipolar disorder and emotional incontinence are present in a consistent percentage of patients. The direct consequence is a negative effect on patient's quality of life and caregiver burden, with different degrees. Other psychiatric manifestations, such as psychotic disorders, adjustment disorders, personality disorders, drug addiction, and abuse of substances, are less frequent. The pathogenesis of psychiatric disturbances in CADASIL is incompletely understood, but, similarly to other cerebrovascular diseases, may depend on the damage of the cortical–subcortical circuits, leading to the consideration of mood disorders in CADASIL under the concept of "vascular depression". In conclusion, CADASIL symptoms may be highly variable, usually starting in adulthood, but also reported in older age. The wide clinical spectrum includes complex migraine attacks with prominent aura, acute confusional states or coma, lacunar strokes, and pure psychiatric or cognitive presentations.

Depression

Depression following a stroke, also referred to as post-stroke depression (PSD), is one of the more frequent complications of stroke, and has significant negative consequences on the recovery of motor and cognitive deficits, as well as the mortality risks associated with stroke, and has negative consequences on the recovery of motor and cognitive deficits, as well as the mortality risks associated with stroke. The prevalence of PSD has ranged from 5 to 63% of patients in several cross-sectional studies, peaking three to six months after a stroke, peaking three to six months after stroke [3, 10, 16]. The systematic review of 51 studies (reported in 96 publications) conducted between 1977 and 2002 by Maree et al. [17] showed that frequencies of depresssion varied considerably across studies, but the pooled estimate was 33% (95% confidence interval, 29–36%) of all stroke survivors experiencing depression.

Major depression and minor depression are the most frequently recognized expression of PSD, and clinical manifestations of PSD are similar to those of idiosyncratic late-onset depression, but psychomotor retardation may be frequently identified. Twenty-five percent of patients hospitalized with an acute stroke develop major depression which is phenomenologically indistinguishable from idiopathic major depression [5]. Moreover, depresssion in stroke also has a bidirectional relationship, as not only are patients with stroke at greater risk of developing depression, but patients with depression have a twofold greater risk of developing a stroke, even after controlling for other risk factors [3, 10].

Anxiety accompanies a large number of neurological disorders, has been observed in many medical conditions, and may be induced by a wide variety of substances [18, 19]. Anxiety that is manifestation of the brain disorder resembles the symptomatology of idiopathic generalized anxiety disorder (GAD), including excessive worry or unwarranted apprehension: shakiness, trembling, and restlessness; shortness of breath; palpitations, excessive sweating; clammy hands, dry mouth, light-headedness, flushes or chills, and frequent urination; and agitation, increased startle response, poor concentration, and irritability [20]. In a study by Starkstein et al. [20] GAD was found in 24% of patients with acute stroke. Most of these patients also had a diagnosis of major depression. GAD alone was found in 6%. In a study by Castillo et al. [21] GAD occurred in 11% of nondepressed stroke patients. In one of our studies [22] anxiety was found in 30% of patients 48 hours after acute stroke and 25% after 15 days of stroke onset. According to study made by Astrom [23] the prevalence of GAD after stroke was 28% in the acute stage, and there was no significant decrease through the 3 years of follow-up. At 1 year, only 23% of the patients with early GAD (0–3 months) had recovered; those not recovered at this follow-up had a high risk of a chronic development of the anxiety disorder. Comorbidity with major depression was high and seemed to impair the prognosis of depression. At the acute stage after stroke, GAD plus depression was associated with left-hemispheric lesion, whereas anxiety alone was associated with right-hemispheric lesion. Cerebral atrophy was associated with both depression and anxiety disorder late but not early after stroke.

Delirium

Delirium, synonymous with the acute confusional state, is a condition of relatively abrupt onset and short duration whose major behavioral characteristics are altered attention. It is acute reversible mental disorders characterized by confessional state with disorientation for time or place [18]. Other behavioral abnormalities frequently coexist including mood and emotional alterations, illusions, hallucinations with increased or decreased psychomotor activity.

There are different reports on frequency of delirium in acute stroke, from 24 to 48%, and it is more frequent in hemorrhagic then ischemic stoke [10, 16, 24]. Delirium is not stable state. The level of consciousness may be reduced or may fluctuate between drowsiness and hypervigilance, but the patient is unable to maintain attention for any substantial period of time. The principal effort in the management of the patient in delirium is directed at identifying and treating the underlying disease process [10].

Post-stroke Language Disorders

Post-stroke language disorders are frequent and include aphasia, alexia, agraphia, and acalculia. There are different definitions of **aphasias**, but the most widely accepted neurologic and/or neuropsychological definition is that aphasia is a loss or impairment of verbal communication, which occurs as a consequence of brain dysfunction. It manifests as impairment of almost all verbal abilities, e.g., abnormal verbal expression, difficulties in understanding spoken or written language, repetition, naming, reading, and writing. During history, many classifications of aphasia syndromes were established. For practical use, classification of aphasias according to fluency, comprehension, and abilities of naming it seems to be most suitable (nonfluent aphasias: Broca's, transcortical motor, global and mixed transcortical aphasia; fluent aphasias: anomic, conduction, Wernicke's, transcortical sensory, subcortical aphasia).

Aphasia is a common consequence of left-hemispheric lesion and most common neuropsychological consequence of stroke, with a prevalence of one-third of all stroke patients in acute phase, although there are reports on even higher figures. Many speech impairments have a tendency of spontaneous recovery. Spontaneous recovery is most remarkable in the first three months after stroke onset. Recovery of aphasias caused by ischemic stroke occurs earlier and it is most intensive in the first two weeks. In aphasias caused by hemorrhagic stroke, spontaneous recovery is slower and occurs from the fourth to the eighth week after stroke. The course and outcome of aphasia depend greatly on the type of aphasia. Regardless of the fact that a significant number of aphasias spontaneously improve, it is necessary to start treatment as soon as possible.

The writing and reading disorders in stroke patients (alexias and agraphias) are more frequent than verified on routine examination, not only in less developed but also in large neurologic departments.

Alexia is an acquired type of sensory aphasia where damage to the brain causes the patient to lose the ability to read. It is also called word blindness, text blindness or visual aphasia. Alexia refers to an acquired inability to read due to brain damage and must be distinguished from dyslexia, a developmental abnormality in which the individual is unable to learn to read, and from illiteracy, which reflects a poor educational background. Most aphasics are also alexic, but alexia may occur in the absence of aphasia and may occasionally be the sole disability resulting from specific brain lesions. There are different classifications of alexias. Traditionally, alexias are divided into three categories: pure alexia with agraphia, pure alexia without agraphia, and alexia associated with aphasia ("aphasic alexia").

Agraphia is defined as a disruption of previously intact writing skills by brain damage. Writing involves several elements: language processing, spelling, visual perception, visuospatial orientation for graphic symbols, motor planning, and motor control of writing. A disturbance of any of these processes can impair writing. Agraphia may occur by itself or in association with aphasias, alexia, agnosia and apraxia. Agraphia can also result from "peripheral" involvement of the motor act of writing. Like alexia, agraphia must be distinguished from illiteracy, where writing skills were never developed.

Acalculia is a clinical syndrome of acquired deficits in mathematical calculation, either mentally or with paper and pencil. These language disturbances can be classified differently, but there are three principal types of acalculia: acalculia associated with language disturbances, including number paraphasia, number agraphia, or number alexia; acalculia secondary to visuospatial dysfunction with malalignment of numbers and columns, and primary anarithmetria entailing disruption of the computation process [11].

Neuropsychology of Acute Stroke

Neuropsychology includes both the psychiatric manifestations of neurological illness (primary brain-based disorders) and neurobiology of "idiopathic" psychiatric disorders. Neurological primary brain disorders provoke broad spectrum of brain pathophysiology that causes deficit sin human behavior, and the magnitude of neurobehavioral-related problems is a worldwide health concern. Speech disorders of aphasic type, unilateral neglect, anosognosia (deficit disorders), delirium, and mood disorders (productive disorders) in urgent neurology, first of all in acute phase of stroke are more frequent disorders then it verified in routine exam, not only in the developed and large neurological departments.

Aphasia is common consequence of left-hemispheric lesion and most common neuropsychological consequence of stroke, with prevalence of one-third of all stroke patients in acute phase although exist reports on greater frequency.

Unilateral neglect is a disorder that mostly affects the patient after the lesion of the right hemisphere, mostly caused by a cerebrovascular insult (infarct or hemorrhage affecting a large area—up to two-thirds of the right hemisphere), and in general, the left-side neglect is the most widespread neuropsychological deficit after the lesion of the right cerebral hemisphere. Reports on the incidence of visual neglect vary and they range from 13 to 85%.

Anosognosia is the second place as neuropsychological syndrome of stroke in right hemisphere, characterized by the denial of the motor, visual or cognitive deficit. This syndrome, defined as denial of hemiparesis or hemianopsia, is a common disorder verified in 17–28% of all patients with acute brain stoke [10].

Multiple Sclerosis

Psychiatric syndromes seen in multiple sclerosis (MS) include demoralization, major depression, mania, involuntary emotional expression disorder, cognitive impairment, and psychosis. Demoralization is particularly complex in the context of MS because of the intermittent nature of the condition, which can make it particularly difficult to cope with. Patients usually have more difficulty adapting to acute rather than gradual changes in disease course. They can become increasingly demoralized in a condition that remits, remains quiescent for a while, and then returns, often with more severe symptoms. Several studies suggest that over time many MS patients find it increasingly difficult to adapt psychologically to new episodes and that this can adversely impact their relationships and psychosocial functioning [5, 25].

The high prevalence of depression was recognized in Charcot's early characterization of MS. Review of various studies has indicated the presence of depressive symptoms in approximately 80% of all patients with MS, and estimated lifetime prevalence rates of major depressive disorders to range from 10 to 60% [3, 27, 28]. Diagnosing depression in an MS patient can be difficult because many symptoms such as sleep disorder, fatigue, and apathy overlap with the primary disease. Nevertheless, with careful clinical assessment, depression can be confidently diagnosed. It is a major source of disability and quality of life impairment. Suicidal ideation is fairly prominent in MS patients with the prevalence across the disease of the order of 30% [29].

Treatment of depression in MS should include pharmacotherapy and different types of psychotherapy. There are a few controlled studies that have examined the pharmacotherapy responses of depression in MS patients, and treatment basically should be the same as treatment of idiopathic depression [3]. Euphoria and other manic symptoms have been reported in MS patients back to the days of Charcot. Up to 10% of patients develop euphoria or more severe forms of mania. Additionally, euphoria and mania can be the result of MS treatments, and in particular steroid use.

Cognitive impairment occurs in 40–65% of multiple sclerosis (MS) patients, typically involving complex attention, information processing speed, (episodic) memory and executive functions. It is seen in the subclinical radiologically isolated syndrome, clinically isolated syndrome, and all phases of clinical MS. In pediatric-onset MS cognition is frequently impaired and worsens relatively rapidly. Cognitive impairment often affects personal life and vocational status. Depression, anxiety, and fatigue aggravate symptoms, whereas cognitive reserve partially protects. Cognitive dysfunction correlates to brain magnetic resonance imaging (MRI) lesion volumes and (regional) atrophy, and degree of and increase in MRI abnormalities predict further worsening [30].

Parkinson's Disease

Nowadays, Parkinson's disease (PD) is generally considered a multifaceted disease with a broad spectrum of symptoms and has been associated with cognitive disorders, affective disorders, psychotic phenomena, impulse control disorders, and problematic repetitive behaviors [5]. In an era where the motor symptoms can be relatively well controlled with L-dopa in the early and middle stages of PD, the psychiatric syndromes are often a major source of disability, distress, and quality of life impairment for both patients and caregivers.

Most patients with PD experience some cognitive impairment, with 25% to 40% developing dementia over the course of their illness. Longitudinal studies suggest that the type and severity of cognitive disturbances are stage dependent. In early stages, patients primarily develop problems with memory and information processing, probably as a result of the disease's primary involvement of subcortical structures. In later stages, impairments in cortical functions, such as dyspraxia and amnesia, emerge in many patients [3, 5, 31].

Depressive disturbances are common in PD, with a prevalence of 25% to 50% over the course of the illness. Fewer than half have major depression; most patients have milder forms of depression referred to as dysthymia or subsyndromal depression [3, 5]. In one of our previous studies [32] some depressive symptoms were present in all 35 PD patients, but moderate depression in 22.85% and severe in 45.71%. As in the case of epilepsy and stroke, depression and PD appear to have a bidirectional relationship. That is, not only are patients with PD at greater risk to developing depression, but patients with a depressive disorder have been found to be at greater risk of developing PD [3, 33].

Anxiety is very common in PD but has not been sufficiently studied. Up to 40% of PD patients have anxiety symptoms. Panic disorder is very common, with a prevalence as high as 25%. Panic attacks are fairly typical in their form, in that they are of sudden onset with apprehension and anxiety, associated fears of having a heart attack or dying, and a range of uncomfortable accompanying physical symptoms. The comorbidity of depressive and anxiety disorders in PD is common; most of the time neither occurs alone [5, 34]. According to review paper by Leentjens in 2012 [35] the prevalence and cumulative incidence of psychopathological symptoms are high. The reported prevalence is 17% for major depressive disorder, 34% for anxiety disorder, 17% for apathy, 14% for impulse control disorders, 88% for sleep disturbances, and 60% for sexual problems. The cumulative incidence of hallucinations is 60%. Mild cognitive impairment is present in at least 50% with a cumulative incidence of 66% for dementia after 12 years. All psychopathological syndromes have a strong negative impact on a number of disease parameters, other psychiatric comorbidity, and quality of life. All psychopathological syndromes tend to occur with higher frequency in patients with the hypokinetic rigid type of PD. Other risk factors divide into general and disease-specific risk factors and may vary between the different syndromes.

Huntington Disease

Huntington's disease (HD) is neurodegenerative disease inherited in an autosomal domiant fashion, characterized by progressive movement disorders associated with cognitive and psychiatric symptoms [36, 37]. By the 1980s, the name of the disease becomes known as Huntington's disease, with the recognition of its motor and non-motor signs and symptoms. The underlying genetic defect is an unstable CAG trinucleotide repeat expansion in exon 1 of the HD gene, formerly called IT-15, on the short arm of chromosome 4. A repeat CAG length of 36 and longer is pathogenic and results in the synthesis of an abnormal polyglutamic tract in Huntington's, a widely expressed protein of uncertain function causing accumulation of intracellular protein aggregates, neurotrophic factor deprivation, impairment of energetic metabolism, transcriptional deregulation and, finally, hyperactivation of programmed cell-death mechanism [37].

In most cases, the age of onset of HD is between 35 and 45 years, whereas the mean duration of the disease is 16 years. Different stages of the disease may be described (premanifest, with soft signs, phenoconversion, and manifest), each being characterized by decreasing independence and the need for help caused by a deterioration of motor and cognitive performance and the presence od psychiatric symptoms [38–40].

Clinical pictures of HD comprise motor abnormalities (chorea, dystonia, bradykinesia, and oculomotor dysfunction), cognitive impairment, behavioral problems, and psychiatric disorders [38]. The latter are major constitutions of the clinical spectrum of HD and have a substantial impact on daily functioning, constituting the most distressing aspect (for both patient and relatives) and often the reason for hospitalization.

In the early description of HD, more attention was given to its cognitive features and dementia. Starting in the literature published in the 1990s, many facts of HD have been described including specific psychiatric aspects. Behavioral and psychiatric symptoms (also called prodromal) often precede the manifestation of motor abnormalities of HD. Historical description estimated rates for lifetime prevalence of psychiatric disorders among HD patients vary widely between 33 and 76% [37, 41].

Several recent studies have described neuropsychiatric symptoms including depressed mood, mania, irritability, anxiety, apathy, obsessions-compulsions, and psychosis. Prevalence variation depends on the different methods used to detect and evaluate (e.g., interview, rating scales, self-report questionnaires) [37]. The most frequent psychiatric sign occurring in HD patients consists of a depressive symptomatology (DS). The estimated prevalence of depression in HD varies widely, ranging from 9 to 63%, with several studies suggesting rates between 40 and 50% [37].

In his original description of the disease, George Huntington stated that there was "a tendency to insanity and suicide". The suicide rate estimate for HD patients is traditionally 4–6 times higher than in general population. However, recent studies have confirmed the need to distinguish suicide ideation from attempted suicide, as well as considering differently several groups with different characteristics in order not to introduce confounding data [37, 41]. Anxiety has often reported in HD patients, independent of gender (17–61%), both due to the course of the disease and from the neurodegenerative process itself. There are contrasting data regarding the critical stage for anxiety and depression to arise, with some studies identifying stage 2 as the most critical and others suggesting stages 4–5.

The presence of obsessive and compulsive symptoms (OCs) has previously documented in patients with HD as less common than other psychiatric symptoms. However, a study by Marder et al. [42] reported that 22.3% had obsessive and compulsive symptoms as their first clinical visit, suggesting that these symptoms may be more common than previously recognized in this population. These symptoms show an increase with disease progression with a trend similar to the one of depression and anxiety in the stage of the disease [37, 43]. Prevalence of psychotic symptoms (PS), including paranoia and delusional and psychotic states resembling various types of schizophrenia-like psychosis, varies between 3 and 11% [37].

Wilson Disease

Wilson's disease (WD), also named hepatolenticular degeneration, is an autosomal recessive genetic disorder caused by defects of ATP7B gene. This disease occurs sporadically all over the world. It is found in individuals aged 3–80 years, but mainly in children and adolescents, males have a slightly higher risk of developing WD than females, possibly because of differences in estrogen level and iron metabolism. Worldwide prevalence of WD is around 1:30,000, carrier rate is about 0.022, and the gene frequency is about 0.56 [44].

The frequency of distinct neurological features of WD such as dystonia or parkinsonism varies widely in different case series. The presence of classical "wing beating tremor" or "flapping temor" in combination with dysarthria strongly suggests the diagnosis of WD. However, any of the other, more common forms of tremors such as rest, action, or intention tremor can occur as well. The most common form of tremor in WD is an irregular, and somewhat jerky, dystonic termor. Dystonia is present in at least a third of all patients with a neurological presentation of WD and can be generalized, segmental, multifocal or focal [45].

Psychiatric symptoms can occur in both untreated and treated WD patients. According to one recent literature review, 20% will have seen a psychiatric before a formal diagnosis of WD was reached. The average time between the onset of psychiatric symptoms and the diagnosis of WD was 864 days for WD patients in whom psychiatric symptoms preceded neurological or hepatic involvement. The most common psychiatric features are abnormal behavior (typically increased irritability or disinhibition), personality changes, anxiety, and depression. Psychosis is considerably less common [46]. The most common psychiatric features are abnormal behavior (typically increased irritability or disinhibition), personality changes, anxiety and depression. Psychosis is considerable less common [47].

Epilepsy

Up to 50% of patients with epilepsy have psychiatric syndromes. Cognitive, mood, anxiety, and psychotic disturbances are most common [5, 18]. Since the epilepsies are heterogeneous and chronic conditions, this complexity is also reflected in the associated psychiatric disturbances. For the most part, psychiatric disturbances have been categorized according to whether they are direct expressions of a seizure, features of a postictal state, or phenomena that occur during the interictal period.

The majority of psychiatric syndromes in epilepsy occur in the interictal period, and thus probably have more to do with the state of the brain in the absence of excessive electrical discharge than with the discharge itself [5].

Depression is the most common psychiatric comorbidity in patients with epilepsy. Prevalence is higher than in a matched population of healthy controls and ranges from 3.9% in patients with controlled epilepsy to 20–55 in patients with recurrent seizures [5]. As in the case of Parkinson's disease and stroke, depression and epilepsy appear to have also a bidirectional relationship [3, 10]. The clinical presentation of depressive disturbances is for the most part typical for idiopathic depression. However, about a third of patients with epilepsy present with atypical features of depression that tends to be intermittent.

They also resemble dysthymia and include anhedonia, fatigue, anxiety, and irritability with less prominent impairments in self-attitude, self-depreciative ideas, or suicidal ideation.

The rate of manic syndromes appear to be higher in epilepsy, and these usually are atypical in presentation and more likely to present with irritability and over-activity than idiopathic bipolar disorder, which itself does not appear to be more prevalent in epilepsy relative to the general population. This has led to the belief that epilepsy-associated brain damage is a major component in the occurrence of mania and temporal lobe epilepsy.

The prevalence of psychotic symptoms in interictal periods is on the order of 5% to 7% in patients with epilepsy. In patients with temporal lobe epilepsy, these disturbances are often schizophrenia-like in their presentation. Paranoid or persecutory delusions and both visual and auditory hallucinations have been reported. Also "negative symptoms" of schizophrenia such as motivation, apathy, flattened affect, and disorganized behavior have been reported in association with delusions and hallucinations. This has given rise to the hypothesis of the "schizophrenia-like psychoses of epilepsy" which remains controversial [5, 48].

References

1. Moriarty J. Psychiatric disorders in neurology. J Neurol Neurosurg Psychiatry. 2007;78:331.
2. Jefferies K, Owino A, Rickards H, Agrawal N. Psychiatric disorders in inpatients on a neurology ward: estimate of prevalence and usefulness of screening questionnaires. Neurol Neurosurg Psychiatry. 2007;78:414–6.
3. Kanner MK. Depression in neurological disorders. Wilingham: Cambridge Medical Communication Ltd; 2005.
4. Sinanović O. Psychiatric disorders in neurology. Pychiat Danub. 2012;24(Suppl. 3):331–5.
5. Lyketsos GC, Kozauer N, Rabins VP. Psychiatric manifestations of neurological disease: where are we headed? Dialogues Clin Neurosci. 2007;9:111–24.
6. Devasenapathy A, Hachinski V. Cerebrovascular disease. In: Rizzo M, Eslinger JP, editors. Behavioral neurology and neuropsychology. Philadelphia: W.B. Saunders Company; 2004. p. 597–613.
7. Morris PL, Robinson RG, Andrezejewski P, Samuels J, Price TR. Association of depression with 10-year poststroke mortality. Am J Psychiatry. 1993;150:124–9.
8. Cummings JL, Arciniegas DB, Brooks BR, et al. Defining and diagnosing involuntary emotional expression disorder. CNS Spectr. 2006;11:1–7.
9. Roman GC, Sachdev P, Royall DR, et al. Vascular cognitive disorder: a new diagnsotic category updating vascular cognitive impairment and vascular dementia. J Neurol So. 2004;226:81–7.
10. Sinanović O. Neuropsychology of acute stroke. Psychiat Danub. 2010;22:278–81.
11. Sinanović O, Mrkonjić Z, Zukić S, Vidović M, Imamović K. Post-stroke language disorders. Acta Clin Croat. 2011;50:77–92.
12. Mijajlović MD, Pavlović A, Brainin M, Heiss W-D, Quinn TJ, Ihle-Hansen HB, Hermann DM, Assayag EB, Richard E, Thiel A, Kliper E, Shin Y-I, Kim Y-H, Choi SH, Jung S, Lee Y-B, Sinanović O, Levine DL, Schlesinger I, Mead G, Milošević V, Leys D, Hagberg G, Ursin MH, Teuschl Y, Prokopenko S, Mozheyko E, Bezdenezhnykh A, Matz K, Aleksić V, Muresanu DF, Korczyn AD, Bornstein NM. BMC Med. 2017;15:6–12.
13. Seshadri S, Wolf PA. Lifetime risk of stroke and dementia: current concepts, and estimates from the Framingham study. Lancet. 2007;3:1106–14.
14. Narayan SK, Gorman G, Kalaria RN, Ford GA, Chinnery PF. The minimum prevalence of CADASIL in northeast England. Neurology. 2012;78(13):1025–7.
15. Di Donato I, Bianchi S, De Stefano N, Dichgans M, Dotti MT, Duering M, Jouvent E, Korczyn AD, Lesnik-Oberstein SAJ, Malandrini A, Markus HS, Pantoni L, Penco S, Rufa A, Sinanović O, Stojanov D, Federico A. Cerebral Autosomal Dominant Arteriopathy with Subcortical Infarcts and Leukoencephalopathy (CADASIL) as a model of small vessel disease: update on clinical, diagnostic, and management aspects. BMC Med. 2017;15:41.
16. Robinson RG. Posstroke depression. Prevalence, diagnosis, treatment and disease progression. Biol Psychiatry 2003;54:376–87.
17. Maree LH, Yapa C, Parag V, Anderson CS. Frequency of depression after stroke. A systematic review of observational studies. Stroke 2005;36:1330–43.
18. Cummings JL, Mege MS. Neuropsychiatry and bihevioral neuroscience. Oxford: Oxford University Press; 2003.

19. Cunnings J. Neuropsychiatry. In: Simpson MG (ed). New York: Impact Communications, Inc; 1995. p. 109–136.
20. Starkstein SE, Cohen BS, Fedoroff P, Parikh RM, Price TR, Robinson RG. Relationship between anxiety disorders and depressive disorders in patients with cerebrovascular injury. Arch Gen Psychiatry. 1990;47:246–51.
21. Castillo S, Starkstein SE, Fedoroff P, Price TR, Robinson RG. Generalized anxiety after stroke. J Nerv Ment Dis. 1993;181:100–6.
22. Ibrahimagić OĆ, Sinanović O, Smajlović D. Anxiety in acute phase of ischemic stroke and myocardial infarction. Med Arh. 2005;59:366–9.
23. Astrom M. Generalized anxiety disorder in stroke patients—a 3-year longitudinal study. Stroke. 1996;27:270–5.
24. Sinanović O, Vidović M, Smajlović Dž. Najčešći neuropsihološki poremećaji u akutnom cerebrovaskularnom inzultu. Liječ Vjesn. 2006;128(Supl 6):20–21.
25. Mohr DC, Dick LP, Russo D, et al. The psychosocial impact of multiple sclerosis: exploring the patient's perspective. Health Psychol. 1999;18:376–82.
26. Dostović Z, Smajlović Dž, Sinanović O, Vidović M. Duration of delirium in acute stage of stroke. Acta Clin Croat. 2008;48:13–17.
27. Minden SL, Schiffer RB. Affective disorders in multiple sclerosis, review and recommendations for clinical research. Arch Neurol. 1990;47:98–104.
28. Siegert RJ, Abernethy DA. Depression in multiple sclerosis: a review. J Neurol Neurosurg Psychiatry. 2005;76:469–75.
29. Feinstein A. An examination of suicidal intent in patients with multiple sclerosis. Neurology. 2002;59:674–8.
30. Jongen PJ, Ter Horst AT, Brands AM. Cognitive impairment in multiple sclerosis. Minerva Med. 2012;103:73–96.
31. Marsh L, Berk A. Neuropsychiatric aspects of Parkinson's disease. Recent advances Curr Psychiatry Rep. 2003;5:68–76.
32. Sinanović O, Hudić J, Ibrahimagić O. Dementia and depression in Parkinson's disease. Parkinsonism Relat Disord. 2007;13(Suppl 2):S53.
33. Nilson FM, Kissing LV, Bowling TG. Increased risk of developing Parkinson's disease for patients with major affective disorder a register study. Acta Psychiatr Scand. 2001;104:380–6.
34. Schneider F, Althaus A, Backes V, Dodel R. Psychiatric symptoms in Parkinson's disease. Eur Arch Psychiatry Clin Neurosci. 2008;258(Suppl 5):55–9.
35. Leentjens AFG. Epidemiology of psychiatric symptoms in Parkinson's disease. Adv Biol Psychiatry. 2012;27:1–12.
36. Van Duijn E, Kingma EM, Timman R, Zitman FG, Tibben A, Ros RA, van der Mast RC. Cros-sectional study on prevalences of psychiatric disorders in utation carriers of Huntington's disease compared with mutation-negative first-degree relatives. J Clin Psychiatry. 2008;69:1804–10.
37. Paoli RA, Botturi A, Ciammola A, Silani V, Prunas C, Lucchiari C, Zugano E, Calet E. Neuropsychiatric burden in Huntington's disease. Brain Sci. 2017;7(6):67. https://doi.org/10.3390/brainsci7060067.
38. Bates GP, Harper PJL. Huntington's disease. Oxford: Oxford University Press; 2002.
39. Terzić R, Tupković E, Logar N, Šehić A, Đuričić E, Sinanović O, Peterlin B. Primjena DNA testa u dijagnostici Huntingtonove bolesti. Med Arh. 2002;56(4):187–9.
40. SinanovićO. Etička kompleksnost genetskog savjetovanja porodice i bolesnika koji boluje og Huntingtonove bolesti U: Valjan V (urednik). Integrativna bioetika i interkulturalnost (Zbornik radova Drugog međunardonog bioetičkog simpozija u Bosni i Hercegovini, Sarajevo, od 23. do 24. svibnja 2008). Sarajevo: Bioetičko društvo BiH, 2009. p. 177–183.
41. Cummings JL. Behavioral and psychiatric symptoms associated with Huntington's disease. In: Winer WJ, Lang AE, editors. Behavioral neurology of movement disorders. New York: Raven Press; 1995. p. 179–86.

42. Marder K, Zhao H, Myers RH, Cudkowicz M, Kayson E, Kieburtz K, Orme C, Paulsen J, Penney JB, Siemenrs E, et al. Rate of functionl decline in Huntington's disease. Neurology 2000; 54: 452–8.
43. Orth M, Handley OJ, Schwenke C, Dunnett SB, Craufurd D, Ho AK, Wid E, Tabrizi SJ, Landwehrmyer GB. The investigators of the European Huntington's disease network observing Huntington's disease: European Huntington's disease Network's REGISTRY. PLoS Curr. 2010. http://dx.doi.org/10.1136/jnnp.2010.209668.
44. Litwin T, Gromadzka G, Czlonkowska A. Gender differences in Wilson's disease. J Neurol Sci. 2012;312:31–5.
45. Svetel M, Kozić D, Stefanova E, Semnic R, Dragašević N, Kostić VS. Dystonia in Wilson's disease. Mov Disord. 2001;16(4):719–23.
46. Zimbrean PC, Schilsky ML. Psychiatric aspects of Wilson disease: a review. Gen Hosp Psychiatry. 2014;36(1):53–62.
47. Bandmann O, Weiss KH, Kaler SG. Wilson's disease and other neurological copper disorders. Lancet Neurl. 2015;14(1):103–13.
48. Lyketsos CG, Stoline AM, Longstreet P, Lesser R, Fisher R, Folstein MF. Mania in temporal lobe epilepsy. Neuropsychiat Neuropsychol Behav Neurol. 1993;6:19–25.

Huntington's Disease

Miroslav Cuturic

History and Epidemiology

In his seminal 1872 paper, George Huntington provided a detailed description of a hereditary illness that emerges in mid-life, manifested by progressive chorea and clinical deterioration [1]. Subsequently, William Osler recognized the heredity of the illness as a clear example of an autosomal dominant Mendelian inheritance pattern [2]. George Huntington's precise, articulate, and crisp description of this illness led to its designation as Huntington's chorea and, subsequently, as Huntington's disease (HD). The responsible gene mutation was identified on the short arm of chromosome 4 in 1993 [3].

Although Huntington believed that the disease was restricted mainly to his native Long Island, today we know that HD is a globally widespread disorder with an approximate worldwide prevalence of 5–10 per 100,000 population, notwithstanding some regional and ethnic variability [4]. There is a notably lower prevalence of HD in Asia as compared to Europe, North America, and Australia [5], while some countries have distinctively low prevalence rates, such as Japan and Finland, with respective 0.1 and 0.5 HD cases per 100,000 population [6–8]. In the US, HD has been designated as an orphan disorder, with 25–35,000 individuals having clinical illness and 2–3 times as many pre-symptomatic gene carriers [4, 9].

M. Cuturic (✉)
Department of Neurology, University of South Carolina School of Medicine, 8 Medical Park, Suite 420, Columbia, SC 29203, USA
e-mail: miroslav.cuturic@uscmed.sc.edu

© Springer Nature Switzerland AG 2020
V. Demarin (ed.), *Mind and Brain*, https://doi.org/10.1007/978-3-030-38606-1_8

Natural History

Huntington's disease (HD) is a complex neurodegenerative disorder with a highly penetrant autosomal dominant inheritance pattern, with both sexes having a 50% chance of inheriting the genetic defect from the affected parent. The genetic defect consists of CAG trinucleotide expansion on the short arm of chromosome 4, which translates into polyglutamine chain expansion in the mutant huntingtin protein, resulting in abnormal protein aggregation and neurodegeneration. The illness usually emerges in mid-life, around the mean age of 40 years. The course is manifested by progressive motor, cognitive, and psychiatric deterioration. Juvenile-onset, before the age of 20 years, is seen in up to 10% of cases [10, 11].

In affected adults, chorea is the most notable feature and is often seen as the initial symptom of the disease. However, up to several years prior to the onset of chorea, subtle personality changes, psychological, and motor deficits can be identified on a detailed examination in individuals otherwise deemed asymptomatic [12, 13]. Neuroimaging studies have demonstrated subtle but definite structural changes in pre-symptomatic patients, particularly progressive atrophic changes in the striatum and caudate nuclei [14, 15]. The progression of motor symptoms is associated with intellectual decline and psychiatric disturbances, as a result of neurodegeneration. The rate of progression and duration of the illness may vary, but the majority of patients will survive 10–20 years after onset [16]. With the progression of HD, immobility and dystonia become more prominent, together with dysphagia, which directly contributes to weight loss and aspiration pneumonia, a common cause of death [17]. Terminally, pneumonia and cardiovascular disease are cited as common causes of death [18].

Juvenile HD usually presents with a hypokinetic form of the disease manifested primarily by bradykinesia, dystonia, and rigidity. Chorea is not a prominent feature, while myoclonus and epilepsy are more frequently seen than in the adult form [11, 19]. In affected children, initial symptoms may present as personality changes, attention and concentration disturbances, followed by a decline in cognitive function manifested by loss of previously achieved milestones and skills, as well as a steep decline in school performance and behavioral disturbances. Juvenile-onset HD has a higher risk of rapid progression and is more likely to be inherited from an affected father [19–21].

There is a clear relationship between the number of CAG repeats within the huntingtin gene and the expression of the disease. Individuals with 40 or more repeats will have a full expression of HD, with the progressive motor, cognitive, and psychiatric symptoms, while individuals carrying 36–39 repeats will have reduced penetrance, and attenuated or incomplete expression of the illness. Offspring in both of these groups carry a 50% risk of inheriting the illness. Individuals with 27–35 repeats will not develop HD, but their offspring will be at risk due to the phenomenon of meiotic instability [21, 22]. Individuals with less than 27 repeats will not develop HD, and their offspring will not be at risk. There is also a confirmed inverse correlation between age at onset of HD and CAG repeat length within the abnormal Huntingtin gene [23–27]. This correlation accounts for only about 50% of the variance in age at onset. Therefore, CAG repeat length alone is not a sufficient predictor of onset [23]. There is also familial influence on age of onset which is independent of the effect of the CAG repeat expansion, suggesting a contribution from additional modifiers [28]. The relationship between CAG repeat length and the rate of clinical progression of illness remains unsettled.

Pathology and Pathophysiology

There is a selective pattern of neuronal vulnerability to huntingtin protein aggregation, with a particular susceptibility and loss of medium spiny neurons that use gamma-aminobutyric acid (GABA) as their neurotransmitter [29, 30]. In the early stages of HD, medium spiny neurons projecting into the external segment of the globus pallidus (GPe) are more prominently lost in comparison to those projecting into the internal segment of the globus pallidus (GPi) [31]. This differential loss of striatal projections has been postulated to result in imbalanced activity in the so-called direct and indirect pathways, causing chorea [31, 32]. More balanced loss of neurons projecting to both, GPi and GPe may result in a rigid-akinetic variant of HD, such as seen in juvenile cases [33]. Therefore, in the early stages, increased dopamine neurotransmission results in hyperkinetic movements, while, in the late stages, dopamine deficits produce hypokinesia [34]. During the long process of progression, several types of neurons vulnerable in HD undergo proliferative and degenerative alterations [35, 36]. At the point when neurons become unable to compensate for the ongoing cellular stress, they degenerate in a process similar to apoptosis, finally resulting in cell death [37].

The severity of atrophy in the striatum, as well as in the cortex and thalamus, correlates with the clinical progression of the disease [38, 39]. There is a 15–30% loss in brain weight that occurs in the course of HD [39, 40]. The most prominent pathological changes in HD are seen in the neostriatum, manifested particularly by atrophy of the caudate nucleus and putamen, with associated neuronal loss, astrogliosis, as well as reactive microgliosis [41, 42]. Neuropathological changes may precede clinical onset by several years and maybe manifested by huntingtin protein aggregation, striatal atrophy, neuronal loss, and oligodendrogliosis [43]. In the process of globus pallidus atrophy, GPi and GPe may lose more than 50% of volume and 40% of neurons, while glia increases in concentration, as well as in absolute number [40]. Atrophy and gliosis in HD have also been described in the substantia nigra, including both the pars compacta and pars reticulata, with a cross-sectional area loss up to 40% [36, 44]. Neuronal loss has also been found within the thalamic and subthalamic nuclei but has not been studied extensively [40, 45]. In the hypothalamus, the significant neuronal loss has been described in the supraoptic nucleus, lateral hypothalamic, and lateral tuberal nuclei, which has been postulated to play a role in cachexia in HD [46, 47]. Generalized cortical atrophy is often noted at autopsy, and accounts for most of the loss in brain mass associated with HD [39, 40]. The loss of neurons and volume is most prominent in cortical layers III, V, and VI [48, 49]. The mean neuronal loss in the entire cortical hemisphere maybe as high as 33%, while astrocyte and oligodendrocyte concentrations may significantly increase, particularly in layers III–VI [50].

In the cells of individuals with Huntington's disease, both mutant and normal huntingtin proteins are present [51–53]. The function of normal, wild-type, huntingtin is not fully elucidated. Within the cell, normal huntingtin is associated with dendritic microtubules, as well as organelles such as mitochondria, transport vesicles, synaptic vesicles, and portions of the endocytic system [54–57]. Such associations and distribution allude to its role in the function of these organelles. The expanded CAG trinucleotide repeat sequence in the huntingtin gene translates into an elongated polyglutamine chain that results in an abnormal conformational change in the protein [51, 58]. In the process of misfolding, the polyglutamine sequence bonds within itself and with other molecules [59]. In addition, mutant huntingtin undergoes proteolytic cleavage producing fragments that form macromolecular aggregates among themselves and with other proteins, visible in the cytoplasm, processes and nuclei of neurons [54]. These aggregates can be found throughout the brain particularly in the cortical regions, and less so in the striatum [55, 60]. Relatively high concentrations of huntingtin aggregates are found in dendrites and axons, with lower concentrations in neural cell bodies and nuclei [55, 56].

Aggregation of huntingtin fragments is dependent on the length of its glutamine repeats, occurring when 39 or more molecules in the polyglutamine chain are present [58, 61, 62]. However, the role of aggregation directly causing neurodegeneration has been questioned due to findings of a relatively low concentration of aggregates in the striatum compared to other areas of the brain [55, 56]. In addition, the vulnerable medium spiny neurons are less prone to have huntingtin aggregates than the resistant striatal interneurons [63]. There is a broader agreement that the effects of mutant huntingtin are the result of its abnormal interactions with proteins involved in signal transduction and metabolism, endocytosis and endosome transport, as well as proteins involved in intracellular transport of organelles [64–72]. Additionally, mutant huntingtin protein interacts with proteins associated with gene transcription, as well as proteolytic enzymes, such as caspases and calpains, which have been implicated in the production of toxic fragments of the mutant protein [73].

In addition to the above-mentioned protein aggregation, cleavage and degradation, there are multiple pathways proposed to play a role in the etiology of neurotoxicity in HD. These include transcriptional dysregulation [74], mitochondrial energy dysfunction [75], glutamate and dopamine excitotoxicity [76, 77], brain-derived neurotrophic factor (BDNF) deficit [78], axonal transport impairment [79], autophagy, and immune system-mediated neuroinflammation [80–82]. Despite significant gains in the understanding of the underlying pathophysiological mechanisms in HD, we still have not materialized these advancements into the development of viable disease-modifying therapies.

Neuroimaging

Neuroimaging is not routinely used in the diagnosis of HD, as it has been rendered obsolete by genetic testing. However, novel imaging techniques provide new opportunities in HD research, as they enable us to study brain changes in vivo and follow morphological changes that take place as the disease progresses. High-resolution Magnetic Resonance Imaging (MRI) has been used to obtain accurate measurements of brain atrophy. In addition to atrophy and volume loss in the striatum and basal ganglia, more recent studies have demonstrated regional cortical thinning in the frontal, parietal, posterior temporal, parahippocampal, and occipital regions, some of which have been found even in pre-symptomatic individuals [83–85]. Selective atrophic changes have been shown to correlate with total functional capacity and duration of symptoms [85].

Besides striatal dysfunction, functional MRI imaging studies have reported a variable pattern of increased and decreased activation in cortical regions in both pre-clinical and clinically manifest HD gene mutation carriers. Beyond regional brain activation changes, evidence from functional and diffusion-weighted MRI further suggests disrupted connectivity between corticocortical and corticostriatal areas. However, substantial inconsistencies with respect to structural and functional changes have been reported [86].

Use of Positron Emission Tomography (PET) imaging for evaluation of brain metabolism, postsynaptic dopaminergic function and phosphodiesterase 10A has shown promise in assessing disease progression. However, no single technique may be currently considered an optimal biomarker, and an integrative multimodal imaging approach, combining different techniques, may be needed for monitoring potential neuroprotective and preventive treatment in HD [87].

Clinical Features

Although HD is the result of a single-gene mutation, with a well-defined pathogenic protein, its clinical picture is strikingly complex, manifested by a unique combination of motor, cognitive, and psychiatric symptoms in each individual case. Therefore, each patient should be evaluated in a comprehensive manner to define their specific needs and treatment plan. However, for the purposes of this chapter, we will discuss the features of each symptom set separately.

Motor Symptoms

Movement disorders are the hallmark of HD, with chorea being its most recognizable symptom. Although the term "chorea" is derived from the Greek verb meaning "to dance," there is very little compatibility with this term in Huntington's chorea, as its clinical appearance lacks any symmetry, rhythmicity or graciousness. Chorea in HD is manifested by involuntary movements which are sudden, irregular, asynchronous, purposeless, but frequently "masked up" into semi-purposeful movements. It usually emerges in the distal muscles of the extremities, but with the progression of the disease, it gradually spreads to involve more proximal muscles in the extremities, face, neck, and paraspinal muscles. With further progression, chorea increases in frequency, duration, and amplitude, sometimes to ballistic proportions [88]. Although chorea is the most dramatic symptom, the affected individual may be able to function quite well, even with relatively prominent movements [89]. Nonetheless, in very advanced cases, chorea will interfere with self-care and activities of daily living. Finally, in the later stages of the disease, chorea settles down and gradually yields to dystonia, rigidity, and contractures [90].

Although not as noticeable as chorea, dystonia also presents as a relatively early motor feature of HD, adding to the choreoathetoid and writhing appearance of the involuntary movements. Dystonia is frequently manifested by repetitive, abnormal muscle contraction, sometimes with a twisting component. With subsequent progression, dystonia involves the limbs, neck, and trunk resulting in abnormal and prolonged posturing. In the terminal stages of the disease, dystonia, bradykinesia, rigidity, and contractures dominate the clinical picture [90]. Particularly in juvenile-onset HD (Westphal variant), dystonia, rigidity, and bradykinesia are prominent from the onset of the disease and dominate throughout the course of the illness, while chorea is not as prominent [19, 91]. In general, dystonia may have a more detrimental effect on daily functioning than chorea does, as it contributes to postural instability, dysarthria, and dysphagia resulting in falls, communication difficulties, and aspiration pneumonia. Myoclonus maybe seen in the adult-onset form, but it is much more frequently encountered in the juvenile variant of HD [91]. Additionally, some HD patients may exhibit utterances, vocalizations and tics akin to Tourette's syndrome [92].

In addition to involuntary movements, progressive impairment of voluntary motor control is a core feature of HD, occurring relatively early in the disease. Particularly, voluntary eye movement abnormalities are some of the earliest signs of HD. Interrupted smooth pursuit, slow initiation, and impaired coordination of voluntary saccades are typical features, together with restrictions in the range of eye movements, particularly in a vertical plane. Even in presumably pre-symptomatic gene carriers, a detailed exam may reveal significantly more abnormalities of ocular function than in gene negative individuals [93]. Worsening in manual dexterity is another early sign of HD, manifested in the examination as slowness in finger tapping and rapid alternating movements of the hands. Motor impersistence is manifested by the inability to maintain voluntary motor contraction, which further contributes to difficulties with manual tasks. Clinical exam for motor impersistence reveals an inability to maintain prolonged tongue protrusion and "milk-maid's grip." It has been proposed that depressed synaptic transmission plays a role in motor impersistence [94]. Upper motor neuron signs also may occur in the course of HD, such as spasticity, clonus, and extensor plantar responses. Impairment of voluntary motor control correlates with a disability and functional decline perhaps even more so than involuntary movement disorder [89]. Unfortunately, there are no effective medications to improve voluntary motor control.

The compounding effects of involuntary movements and impaired voluntary motor control result in a multi-faceted gait disorder in more advanced stages of HD. Progressive worsening of dystonia, ballistic chorea, motor impersistence, and voluntary eye movements all contribute to incoordination, impaired ambulation with a propensity towards falls and injuries, resulting in wheelchair dependence [95].

Cognitive Disorder

Although movement disorders and chorea are the most recognized symptoms of HD, cognitive impairment is a much more disabling and distressing manifestation of the disease and presents the greatest burden to the patients and their families [96]. Dementia in HD is classified as "subcortical," as it lacks typical cortical deficits such as aphasia, amnesia, and agnosia, typically seen in Alzheimer's disease [97, 98]. The cognitive disorder in HD consists of bradyphrenia, impairments in attention, sequencing, executive function, perceptual skills, as well as learning and memory impairments [97, 99, 100]. Registration and immediate recall are relatively spared, while retrieval of recent and remote memories are impaired [97, 99, 100]. Explicit sequence learning appears to be more affected than implicit, both in pre-manifest and manifest individuals [101]. For routine monitoring of cognitive status, most HD clinics utilize the Unified Huntington Disease Rating Scale (UHDRS), which, in addition to motor and behavioral scales, also incorporates reliable indicators of cognitive decline including the Symbol Digit Modality Test, the Stroop Color Word Test, and the Verbal Fluency Subtest of the Multilingual Aphasia Examination [102]. In asymptomatic HD gene carriers, subtle cognitive deficits may be present many years before the onset of motor symptoms [103]. Asymptomatic gene carriers test lower than non-carriers in all portions of the Wechsler Adult Intelligence Scale-Revised (WAIS-R), and there is an inverse correlation between the scores and the number of CAG repeats in the HD gene carriers [104].

Executive functioning and efficiency in HD are affected by changes in the speed of cognitive processing, attention, initiation, planning, and organization, as well as by perseveration, impulsivity and decline of other regulatory processes [105–107]. In HD, patients have difficulty learning new information and retrieving previously learned information, possibly due to impaired speed of processing and organizing information [108]. A change in the speed of cognitive processing is one of the earliest and most sensitive indicators of early HD, as completion of previously ordinary mental tasks becomes more time consuming [105, 108]. The slowing of cognitive processing may result from recruitment of alternate cerebral pathways for cognitive tasks, as an attempt to compensate for deficits in implicit memory [109, 110].

Perceptual problems further compound the cognitive issues from early on, sometimes more than a decade before the onset of more obvious symptoms of HD [103]. More specific perceptual deficits include the inability to recognize the emotions of others communicated by facial expression, perception and estimation of time, spatial perception, and smell identification [111–114]. Lack of awareness of one's own symptoms, actions and feelings may be impaired in as many as one-third of individuals with HD. This is thought to result from interruptions in the pathways between the frontal lobes and the basal ganglia and is felt to be a type of agnosia [115]. Central visual and auditory processing deficits additionally contribute to general perceptual and cognitive impairment [116].

Although typical aphasia or impairment in semantic memory is rarely seen in HD, language and communication are still remarkably affected, primarily by deficits in articulation, initiation, and cognitive processing. Individuals with HD may have difficulties with the integration of thought sequencing, information processing, muscle control, and breathing. Despite impairment in language output, comprehension may be relatively preserved, even in advanced stages of HD [100, 117].

Psychiatric Symptoms

Depression is the most common psychiatric presentation in HD, affecting up to 40% of patients in the course of the illness [118, 119]. The symptoms may start in the pre-motor manifest stage, evidenced by subtle impairment of working memory [97, 120]. Depression in HD has all the symptoms and signs of major depression syndrome, such as sustained low mood, tearfulness, sadness, low self-esteem, loss of appetite, sleep disturbances, feelings of guilt, shame, hopelessness, and helplessness [121]. In severe cases, depression may progress and individuals may develop psychotic symptoms such as delusions and hallucinations or even more profound psychomotor retardation and catatonia. Sometimes it may be difficult to distinguish depression from other symptoms seen with HD, such as apathy, circadian sleep dysregulation, weight loss, impaired attention, and concentration span. Apathy, in particular, is one of the most common symptoms of the disease and is gradually progressive, in synchrony with other symptoms of neurodegeneration, such as motor and cognitive decline [119, 122]. The course of depressive symptoms does not follow the pattern of progression one would expect if the process was directly linked to the pathophysiology of neurodegeneration [88]. On the contrary, depression is most common at early points in the illness, during the period around the initial diagnosis and in the early stage when the impairments begin affecting daily functioning [123]. Subsequently, depressive symptoms seem to decline in prevalence [123]. Nonetheless, the use of strict diagnostic criteria for major depression usually helps to establish an accurate diagnosis.

Symptoms suggestive of bipolar disorder such as elated affect, mania, and agitation, as well as alterations between depressive and manic episodes may develop in approximately 10% of patients [119, 124]. Irritability in HD may be severe, presenting with outbursts of anger and aggression, affecting up to two-thirds of patients in the course of the illness [125]. One should be cautious not to mistakenly diagnose HD-associated dysexecutive syndrome as bipolar disorder, given that some symptoms of the former, such as irritability, impulsivity, disinhibition, and hypersexuality may imitate symptoms of the latter disorder [106].

The risk of suicide was recognized even by Huntington in his publication in 1872 and has remained a significant concern to present. Identified risk factors include depression, single marital status, childlessness, living alone, and family history of suicide [126]. Additional factors that modify the risk include the level of insight, the severity of affective symptoms, and social support [127–129]. Up to 25% of HD patients may attempt suicide in the course of the disease, with a mortality rate of 5–13% [130, 131]. The rate of suicide is higher in pre-symptomatic individuals at risk for HD [18, 130, 132]. Individuals that are undergoing predictive testing for HD may also be at an increased risk for suicide [127, 129]. The risk of suicide should be discussed with patients, their families, and caregivers. As suicide is a preventable outcome of disease, suicidal ideation should be screened for during each clinical visit.

Psychotic features in HD may include a multitude of symptoms, auditory or visual hallucinations, delusions, particularly of the paranoid type, as well as social withdrawal [118, 122, 131]. The lifetime prevalence of psychosis in HD is between 1 and 15%. However, additional familial modifiers may affect the incidence [119, 122, 133]. In cases when psychiatric symptoms occur before motor symptoms, patients may be mistakenly diagnosed with schizophrenia [134]. In such cases, the subsequent emergence of chorea may be further mistakenly interpreted as neuroleptics related tardive dyskinesia. Therefore, in a psychiatric setting, a high index of suspicion for HD is advised in cases of progressive chorea, in association with an uncertain or unreliable family history. In most such cases, clarifying the family history and monitoring for other signs of HD is sufficient to confirm the diagnosis. However, in certain instances, the issue can only be settled by genetic testing.

Although recent theories have implicated the basal ganglia and frontal lobes in the development of obsessive-compulsive disorder (OCD), HD patients rarely develop the full syndrome. Nonetheless, some HD patients may develop a preoccupation with misophobia, contamination, or may engage in excessive checking and rechecking routines. Some individuals may become fixated on a perceived need or a prior unresolved issue [135]. If the symptoms become so severe as to interfere with the quality of life, a treatment usually applied for OCD may be indicated.

Sleep disturbances are common in HD, and they contribute to progression and overall deterioration. Polysomnographic sleep patterns in HD have been studied sporadically in small groups of patients, providing variable results. However, more recent studies have confirmed delayed nocturnal sleep onset latency as one of the earliest sleep-related findings, with a virtual absence of nocturnal respiratory disturbances in early HD [136]. Polysomnographic studies correlated delayed sleep onset, awakenings, and reduction of slow-wave sleep with the severity of motor symptoms, duration of illness, and degree of caudate atrophy on computed tomography (CT) [137].

Personality changes are hard to categorize but are clearly evident very early in HD. There are various terms used to define them, such as organic personality disorder, dysexecutive syndrome or frontal lobe syndrome, as there is no specific diagnosis to encompass them. The earliest personality changes are often manifested by anxiety, irritability or apathy [138]. Overall, personality changes seem to be widespread in HD, perhaps more common than depression, and sometimes may precede the onset of cognitive and motor symptoms by many years [12, 139, 118].

Diagnosis

HD should be considered in the differential diagnosis of any disorder presenting with a combination of symptoms that includes chorea, dementia, and psychiatric disturbances. The diagnosis is relatively easy to establish in individuals with typical clinical features and a positive family history of HD. Non-inherited disorders that can present with chorea such as thyrotoxicosis, cerebral lupus, cerebrovascular disease, polycythemia or tardive dyskinesia usually can be excluded by routine laboratory tests and by their clinical course. The recommended basic laboratory workup would include a blood count, with a smear for acanthocytes, sedimentation rate, metabolic panel, thyroid parameters, vitamin B12 level, and autoantibodies including lupus anticoagulant, antinuclear, anticardiolipin, antistreptolysin, and anti-DNase-B antibodies [140]. As there are viable treatment options for it, Wilson's disease should always be included in the differential diagnosis of movement disorders and screened for by serum ceruloplasmin levels [141].

There are several familial disorders with clinical features that may overlap with HD, such as X-linked McLeod neuroacanthocytosis syndrome, as well as autosomal recessive disorders, including Chorea-acanthocytosis, Spinocerebellar ataxia type 17, and Dentatorubral-pallidoluysian atrophy [142–145]. Even a very detailed family history may not be able to distinguish the latter disorders from HD. There are also at least two disorders with an autosomal dominant inheritance pattern and clinical features very similar to HD, that should be taken into consideration: Huntington disease-like1, a slowly progressive prion disease [146, 147], and Huntington disease-like 2, that is clinically indistinguishable from HD, but is much more prevalent in individuals of African descent [148]. Negative family history does not necessarily exclude a diagnosis of HD. There are several instances in which HD may be present in the context of negative family history, such as parental or ancestral death before the age of HD expression, an intermediate number of paternal CAG repeats resulting in meiotic instability and expansion in the subsequent generation into HD range, lack of information about family and de novo mutation.

Although clinical evaluation and family history analysis still play a very important role, genetic testing is now considered the gold standard in the diagnosis of HD. When completed in clinically plausible cases, genetic testing has a sensitivity as high as 98.8% and specificity of 100% [149]. Genetic testing can be used for confirmation of the diagnosis in asymptomatic patient, or as predictive testing in an asymptomatic individual at risk for HD. However, the availability of genetic testing in a non-curable disease has raised a myriad of ethical and practical questions [150–152]. Nonetheless, there is a wide consensus that genetic testing should not be a single encounter, but rather a process that involves confidentiality, informed consent, and multidisciplinary supportive counseling, before, during, and after testing and disclosure [153, 154]. The role of psychological and psychiatric counseling is crucial during this process, as the risk of suicide in HD tends to peak around the time of diagnosis [129]. Predictive testing is discouraged in minors, and it is usually pursued only in exceptional circumstances [155]. Despite the availability of the test, only a small proportion, less than 10%, of asymptomatic at-risk individuals decide to undergo predictive testing [154]. Although psychological attributes are similar among individuals who do and do not pursue testing [156], baseline behavioral status has been more strongly associated with the decision to undergo predictive testing than motor symptoms [157]. Following genetic testing, about half of individuals who tested negative for the HD gene had less depression when compared prior to testing, but depression remained the same or worsened in two-thirds of individuals with a positive HD genetic test [157]. In addition, individuals undergoing predictive genetic testing are often concerned about potentially losing their medical insurance as a result of possible subsequent discrimination [156]. Therefore, a structured process of genetic testing for HD that involves psychological and social support, as well as genetic counseling, is justified.

There are several diagnostic options available in the process of family planning for couples affected or at risk for HD. During the initial stages of pregnancy planning, a couple may choose pre-implantation genetic testing in the context of an in vitro fertilization procedure in a specialized center. In this procedure, maternal oocytes are fertilized by the partner's sperm in vitro, and resulting embryos undergo genetic testing prior to implantation. Only embryos without the HD mutation are implanted, assuring that the child born after this procedure will be free of HD [158].

Diagnostic options during pregnancy at risk for HD, include chorionic villus sampling, which can be performed at 8–10 weeks after conception. Amniocentesis may be performed at 14–16 weeks after conception. After the tissue is tested, termination of the pregnancy may be considered by the parents if an HD mutation is found [158]. Non-invasive prenatal diagnosis techniques are also being developed using circulating fetal DNA in the maternal blood to perform testing for the HD mutation during the first trimester of pregnancy [158–160].

Treatment

Currently, there is no disease-modifying therapies or a cure for HD. Therefore, treatment is based largely on lifestyle interventions, supportive management, and symptomatic treatment.

Healthy lifestyle habits may be important not only to symptomatic HD patients but also to asymptomatic carriers. Avoidance of alcohol, drugs, and tobacco is not only good advice for the general population but may also have significant implications in the pre-manifest HD population. A recently completed study on a large cohort of HD patients suggests that the frequent use of tobacco, alcohol, and illicit drugs (including cannabis) may significantly accelerate the onset of motor symptoms in HD, by 2.3, 1.0, and 3.3 years, respectively, with the effects being significantly more prominent in women [161]. Moderate physical activity, such as walking, biking or swimming, is known to have a clear benefit for general health, but particularly in HD, which may help to optimize motor function and hence stabilize motor deficits [162]. In HD, a moderate adherence to the Mediterranean diet has been reported to correlate with a better quality of life, lower comorbidity and less motor impairment [163]. The use of vitamins and dietary supplements is a relatively common practice in the HD community, although such use has not shown any specific benefit to HD patients in clinical trials [164]. Medical practitioners should be vigilant and monitor for signs of overuse of vitamins that may result in toxicity, such as vitamins D, E, K, and A [165].

Due to the complexities of HD, supportive management should not only be aimed at patients but also caregivers and families. Supportive services are most effectively delivered through a comprehensive, multidisciplinary, yet dynamic, program that adjusts to the progressive nature of the disease [166].

In the early stages of the disease, patients and their families can derive a significant benefit from psychological counseling which can help alleviate the stress of genetic testing, manage expectations during the progression of the disease, facilitate effective communication among family members, manage behavioral problems, and monitor for signs of suicide risk. At such time as disease progression affects working capacity, a social worker can assist patients with applications for disability and medical benefits. As the disability progresses, patients and their caregivers benefit from home health service visits for help with daily activities, which in turn may also help prevent caregiver burn-out. With the progression of motor symptoms, physical therapy services may help enhance strength, flexibility, and coordination, help prevent contractures, and also aid in gait reconditioning and fall prevention [167]. Occupational therapy is helpful with training in the use of assistive devices and adapted utensils. Furthermore, periodic speech therapy evaluations are recommended for evaluation and management of swallowing, for prevention of aspiration pneumonia, to improve speech clarity and to provide assistive communication devices when applicable. Multiple observational studies (without control groups) of multidisciplinary rehabilitation in HD demonstrated not only improvements in motor function but also reduction in depression and anxiety [168, 169]. Positive effects on the gray matter have also been shown on neuroimaging, as well as improvement in cognitive function [170]. A living will, advanced medical directives and surrogate decision maker designation should be addressed by a physician or social worker before the patient becomes cognitively unaccountable [171]. In the later stages of the disease, placement in an assisted living or nursing care facility becomes necessary. In terminal stages, enrollment in hospice for comfort care is an appropriate option to consider [172]. Over the years, multidisciplinary clinics have emerged as a destination where patients, caregivers, and affected families can address a multitude of problems and needs at one location. Further initiatives are being pursued with the intent of moving multidisciplinary care from the clinic and outreach to patient's homes as they become immobile with the progression of the illness [173].

Pharmacological therapy in HD is primarily utilized for the treatment of symptoms and does not have any beneficial effect on the progression of the disease. Therefore, initiation and choice of pharmacological therapy for any symptom should be based on the patient's needs and preferences, in corroboration with the caregiver. Often, impairment in patient's perception of their own symptoms may initially result in a delay in treatment [115]. Additionally, the superposition of psychiatric symptoms and cognitive decline affect compliance and increase the risk of complications related to possible incorrect dose intake and interactions. The risks and benefits of medications should be discussed not only with patients but also with caregivers, as well as the possible need, for early supervision with the dispensing of the medications.

Currently, there are no evidence-based treatment recommendations for cognitive decline in HD, as prior clinical trials have not identified any viable pharmacological options. Cholinesterase inhibitors have been proven to be ineffective in designated clinical trials, while preliminary reports from the MITIGATE-HD memantine trial presented at the 2010 Huntington Study Group symposium, suggested worse outcomes for some motor symptoms and only partial improvement of some cognitive measures [174–176]. Apart from general supportive measures, it is important that the physician closely monitor for and correct any toxic metabolic encephalopathies which may result from medications utilized for psychiatric and motor symptoms, as well as other associated conditions in each individual case [177].

The choice and use of pharmacotherapy for psychiatric symptoms are largely based on general psychiatric indications for each medication in the context of the set of psychiatric symptoms in consideration, as there is practically no evidence-based support for its use in HD [176]. As per common practice and expert opinion, selective serotonin reuptake inhibitors (SSRI), as well as serotonin-norepinephrine reuptake inhibitors (SNRI) are the mainstay of pharmacotherapy for depression in HD, primarily due to their favorable side effect profile, safety, and tolerability [178]. Popular SSRIs include fluoxetine, paroxetine, sertraline, citalopram, and escitalopram. Commonly used SRNIs include venlafaxine, duloxetine, and desvenlafaxine. Additional choices may include an atypical antidepressant, such as mirtazapine or bupropion. Tricyclic antidepressants are generally out of favor due to their anticholinergic profile and may worsen hyperkinesia and cognition. Monoamine oxidase inhibitors are largely avoided, due to intolerability. Antidepressants may sometimes exert an initially stimulating effect, resulting in impulsive, disruptive, and self-destructive behaviors. In the United States, all antidepressants carry a black box warning emphasizing that antidepressants may worsen suicidal impulses and behaviors.

For the management of psychotic symptoms with or without depression, the usual choices include new generation neuroleptics such as olanzapine, quetiapine, ziprasidone, aripiprazole, and risperidone [179, 180]. Classic neuroleptics are used less frequently due to a more pronounced side effect profile which may include worsening of cognition, tardive dyskinesia, and dystonia. For the treatment of periodic agitation, irritability or aggressive behaviors, mood stabilizers such as valproic acid, carbamazepine or other anticonvulsant medications are widely utilized, as well as olanzapine, a new generation neuroleptics [179]. Lithium is rarely prescribed due to its potential to worsen involuntary movement, as well as cardiac, endocrine, and metabolic issues. Short-acting benzodiazepines such as lorazepam and alprazolam are preferred choices for acute agitation or anxiety attack treatment. For chronic anxiety, the usual first choices are SSRI's, non-benzodiazepine anxiolytic bupropion or long-acting benzodiazepine clonazepam [178]. Obsessive-compulsive symptoms may respond to SSRI's or a new generation neuroleptic in refractory cases [178]. Although some experts have suggested that stimulants may be beneficial for the treatment of apathy, they should be used with great caution due to their potential to worsen irritability and abuse potential. For sleep disorders, some experts suggest mirtazapine, the benzodiazepine receptor inducer zoldipem, the atypical sedating antidepressant trazodone or new generation neuroleptic quetiapine [176, 178].

There are multiple treatments available for the options of chorea and other motor symptoms in HD. As opposed to the treatment of cognitive and psychiatric symptoms, there are well-defined evidence-based treatments that can be recommended for chorea in HD. The vesicular monoamine transporter-2 (VMAT-2) inhibitors tetrabenazine and deutetrabenazine have demonstrated effectiveness in reducing chorea in HD in well-designed multicenter trials [181, 182]. Both medications received the United States of America Federal Drug Administration (USFDA) approval for treatment of chorea in HD, tetrabenazine in 2008 and deutetrabenazine in 2017. Although both medications demonstrated statistically significant reductions in chorea scores, they have no effect on the natural progression of the illness. Compared to tetrabenazine, the newer medication deutetrabenazine, contains deuterium in its molecule, a naturally occurring non-toxic form of hydrogen. This extends the active metabolites half-lives and minimizes through-to-peak drug concentration fluctuations and peak concentration-related toxicity. In a meta-analysis comparison, deutetrabezine may have less adverse effects than tetrabenazine, particularly a lower rate of psychiatric adverse effects such as irritability, agitated depression, and suicidal ideation [183]. In addition, deutetrabenazine has demonstrated a tendency towards reduction of dystonia [182]. However, head to head comparison studies would be needed to substantiate and clarify such claims. Nonetheless, both medications have a similar side effect profile and carry a USFDA black box warning for potential worsening of suicidal ideation in the context of untreated or inadequately treated depression. Additional contraindications include concomitant MAOI or reserpine treatment and hepatic impairment. Both medications have the same precautions, which include renal impairment, QTc interval prolongation, pregnancy (category C), lactation, and concomitant use of CYP 2D6 inhibitors, such as fluoxetine and paroxetine, which are often used for depression in HD [181, 182].

Both classic and new generation neuroleptics have been used in clinical practice for chorea in HD and have the additional benefit of lessening concomitant psychiatric symptoms. However, neuroleptics also have the potential disadvantage of exacerbating dystonia, tardive dyskinesia, and parkinsonism, with a higher propensity to cause neuroleptic malignant syndrome compared to VMAT-2 medications. Most experts agree that initially, the new generation neuroleptics, should be used due to their better side effect profile, but advanced cases with severe chorea in association with psychosis may require classic neuroleptics. Frequently considered new generation neuroleptics include olanzapine, risperidone, and aripiprazole, while commonly used classic neuroleptics include haloperidol, fluphenazine, and chlorpromazine [184]. The benefits of neuroleptics have been documented mainly in relatively small, open-labeled studies [176]. Additional medications used for the treatment of chorea include amantadine, for which there has been limited concordance in prior studies, and clonazepam, which may also be beneficial for myoclonus, but shares a potential risk of dependency and abuse with other benzodiazepines [178]. In any case, the use of all these medications should be tailored individually to the patient's needs and tolerability.

Pharmacologic treatment of dystonia is often needed in advanced stages of HD, both in the juvenile and adult form and may include benzodiazepines, baclofen, and occasionally dopaminergic anti-parkinsonian medications. Botulinum toxin injections may also be used for focal dystonia [184]. Anticholinergic medications such as benztropine and trihexyphenidyl are best avoided due to potential cognitive side effects, as well as the potential to precipitate delirium.

Experimental Therapeutics and Prospects

Over the last decade and a half, ninety-nine clinical trials have been completed in HD, evaluating 41 compounds and 11 non-pharmacological interventions, including cell therapies, for possible therapeutic effects, with an overall very low success rate at 3.5% [185]. The most significant outcomes were the two USFDA approvals for tetrabenazine and deutetrabenazine for the treatment of chorea.

As the HD gene mutation can be identified decades before disease onset, the ultimate aim of therapy would be to delay the onset of the disease or possibly completely prevent it from emerging. Current HD research is, therefore, focused on finding the most accurate markers of progression in the pre-manifest phase, to enable evaluation of efficacy for potential therapeutic agents prior to the emergence of HD symptoms. Large neuroimaging observational studies have demonstrated that quantitative measurements of the striatum and adjacent brain regions, as well as some cognitive and motor scales, are reliable biomarkers of degenerative progression in pre-manifest HD [186]. Additional markers are being verified including neurofilament light protein in the blood, total tau concentration, and mutant huntingtin protein quantification in the cerebrospinal fluid [187–189].

Correlation between the age at onset of HD and CAG repeat length within the Huntingtin gene, accounts only for approximately 50% of the variance in age at onset, due to the effects of additional genetic modifiers [23, 28]. Hence, there is a remarkable research interest in identifying genetic modifiers that accelerate or delay HD expression as a potential disease-modifying treatment targets [190]. Recently, this has been facilitated by the development of the Genetic Modifiers of Motor Onset Age (GeM MOA) website, where researchers can use single nucleotide polymorphisms as genetic markers within a genome-wide association studies database, in search of genetic HD modifiers [191].

To date, the most promising advances in HD research have been accomplished in the field of gene therapy. In an autosomal dominant disorder such as HD, silencing the mutant gene may potentially have a disease-modifying or even curative effect. The main principles in gene silencing include repression of transcription of DNA information into messenger RNA by using zinc finger proteins, repression of translation of mutant huntingtin by antisense oligonucleotides, and blocking protein translation by RNA interference techniques [160]. Gene silencing techniques have demonstrated a consistent and significant reduction in mutant huntingtin expression in animal HD models [192]. In 2017, a phase 1b–2a clinical trial evaluating the safety of antisense oligonucleotide (ASO) therapy in HD in humans, has been completed. The study utilized ASO designed to inhibit huntingtin messenger RNA and thereby reduce concentrations of mutant huntingtin. It entailed intrathecal bolus application of ASO every four weeks for 4 doses, in patients with early HD. The final report cited a favorable safety profile of this treatment, without the encounter of serious adverse events, with an observed dose-dependent reduction in concentrations of mutant huntingtin in CSF [193]. Open-label extension of this study has continued beyond the completion of 1b–2a phase. In 2018, the pivotal phase 3 trial has been initiated to evaluate the efficacy and safety of this intrathecally administered ASO drug. There are two additional ongoing phase 1b–2a clinical trials evaluating safety and pharmacokinetics of intrathecal application of two distinct allele-specific ASO drugs, designed to lower only mutant huntingtin. All these studies are expected to be completed by late 2020 or early 2021 and HD community is eagerly awaiting the outcomes.

References

1. Huntington G. On chorea. J Neuropsychiatry Clin Neurosci. 2003 Winter;15(1):109–12.
2. Osler W. Historical note on hereditary chorea. In: Browning W, editor. Neurographs. Brooklyn, NY: Albert C. Huntington Publishing; 1908. p. 113–6.
3. The Huntington's disease collaborative research group [no authors listed]. A novel gene containing a trinucleotide repeat that is expanded and unstable on Huntington's disease chromosomes. Cell. 1993;72(6):971–83.
4. Conneally PM. Huntington disease: genetics and epidemiology. Am J Hum Genet. 1984;36(3):506–26.
5. Pringsheim T, Wiltshire K, Day L, Dykeman J, Steeves T, Jette N. The incidence and prevalence of Huntington's disease: a systematic review and meta-analysis. Mov Disord. 2012;27(9):1083–91.
6. Harper PS. The epidemiology of Huntington's disease. Hum Genet. 1992;89(4):365–76.
7. Kanazawa I, Kondo I, Ikeda JE, Ikeda T, Shizu Y, Yoshida M, et al. Huntington's disease genetics. NeuroRx. 2004;1(2):255–62.
8. Sipilä JO, Hietala M, Siitonen A, Päivärinta M, Majamaa K. Epidemiology of Huntington's disease in Finland. Parkinsonism Relat Disord [Internert]. 2015;21(1):46–9. http://sciencedi-rect.com/science/journal/13538020, https://doi.org/10.1016/j.parkreldis.2014.10.025.
9. Tanner CM, Goldman SM. Epidemiology of movement disorders. Curr Opin Neurol. 1994;7(4):340–5.
10. Adams P, Falek A, Arnold J. Huntington disease in Georgia: age at onset. Am J Hum Genet. 1988;43(5):695–704.
11. Reyes Molón L, Yáñez Sáez RM, López-Ibor Alcocer MI. Juvenile Huntington's disease: a case report and literature review. Actas Esp Psiquiatr. 2010;38(5):285–94.
12. Duff K, Paulsen JS, Beglinger LJ, Langbehn DR, Stout JC, Predict-HD Investigators of the Huntington Study Group. Psychiatric symptoms in Huntington's disease before diagnosis: the predict-HD study. Biol Psychiatry. 2007;62(12):1341–6.
13. Kirkwood SC, Siemers E, Bond C, Conneally PM, Christian JC, Foroud T. Confirmation of subtle motor changes among presymptomatic carriers of the Huntington disease gene. Arch Neurol. 2000;57(7):1040–4.
14. Aylward EH, Codori AM, Rosenblatt A, Sherr M, Brandt J, Stine OC, et al. Rate of caudate atrophy in presymptomatic and symptomatic stages of Huntington's disease. Mov Disord. 2000;15(3):552–60.
15. Aylward EH, Harrington DL, Mills JA, Nopoulos PC, Ross CA, Long JD, et al. Regional atrophy associated with cognitive and motor function in prodromal Huntington disease. J Huntingtons Dis [Internet]. 2013;2(4):477–89. http://content.iospress.com/journals/jour-nal-of-huntingtons-disease https://doi.org/10.3233/jhd-130076.
16. Nance M, Paulsen J, Rosenblatt A, Wheelock V. A physician's guide to the management of Huntington's disease. New York: Huntigton's Disease Society of America; 2012. p. 6.
17. Moskowitz CB, Marder K. Palliative care for people with late-stage Huntington's disease. Neurol Clin. 2001;19(4):849–65.
18. Sørensen SA, Fenger K. Causes of death in patients with Huntington's disease and in unaf-fected first degree relatives. J Med Genet. 1992;29(12):911–4.
19. Gonzalez-Alegre P, Afifi AK. Clinical characteristics of childhood-onset (juvenile) Huntington disease: report of 12 patients and review of the literature. J Child Neurol. 2006;21(3):223–9.

20. Bird ED, Caro AJ, Pilling JB. A sex related factor in the inheritance of Huntington's chorea. Ann Hum Genet. 1974;37(3):255–60.

21. Myers RH, Sax DS, Koroshetz WJ, Mastromauro C, Cupples LA, Kiely DK, et al. Factors associated with slow progression in Huntington's disease. Arch Neurol. 1991;48(8):800–4.

22. Lee JM, Ramos EM, Lee JH, Gillis T, Mysore JS, Hayden MR, et al. CAG repeat expansion in Huntington disease determines age at onset in a fully dominant fashion. Neurology [Internet]. 2012;78(10):690–5. http://ovidsp.ovid.com/ovidweb.cgi?T= JS&NEWS=n&CSC=Y&PAGE=toc&D=yrovft&AN=00006114-000000000-00000, https://doi.org/10.1212/wnl.0b013e318249f683.

23. Duyao M, Ambrose C, Myers R, Novelletto A, Persichetti F, Frontali M, et al. Trinucleotide repeat length instability and age of onset in Huntington's disease. Nat Genet. 1993;4(4):387–92.

24. Stine OC, Pleasant N, Franz ML, Abbott MH, Folstein SE, Ross CA. Correlation between the onset age of Huntington's disease and length of the trinucleotide repeat in IT-15. Hum Mol Genet. 1993;2(10):1547–9.

25. Craufurd D, Dodge A. Mutation size and age at onset in Huntington's disease. J Med Genet. 1993;30(12):1008–11.

26. Simpson SA, Davidson MJ, Barron LH. Huntington's disease in Grampian region: correlation of the CAG repeat number and the age of onset of the disease. J Med Genet. 1993;30(12):1014–7.

27. Andrew SE, Goldberg YP, Kremer B, Telenius H, Theilmann J, Adam S, et al. The relationship between trinucleotide (CAG) repeat length and clinical features of Huntington's disease. Nat Genet. 1993;4(4):398–403.

28. Rosenblatt A, Brinkman RR, Liang KY, Almqvist EW, Margolis RL, Huang CY, et al. Familial influence on age of onset among siblings with Huntington disease. Am J Med Genet. 2001;105(5):399–403.

29. Carter CJ. Reduced GABA transaminase activity in the Huntington's disease putamen. Neurosci Lett. 1984;48(3):339–42.

30. Reynolds GP, Pearson SJ. Decreased glutamic acid and increased 5-hydroxytryptamine in Huntington's disease brain. Neurosci Lett. 1987;78(2):233–8.

31. Reiner A, Albin RL, Anderson KD, D'Amato CJ, Penney JB, Young AB. Differential loss of striatal projection neurons in Huntington disease. Proc Natl Acad Sci USA. 1988;85(15):5733–7.

32. Storey E, Beal MF. Neurochemical substrates of rigidity and chorea in Huntington's disease. Brain. 1993;116(5):1201–22.

33. Albin RL, Young AB, Penney JB, Handelin B, Balfour R, Anderson KD, et al. Abnormalities of striatal projection neurons and N-methyl-D-aspartate receptors in presymptomatic Huntington's disease. N Engl J Med. 1990;322(18):1293–8.

34. Cepeda C, Murphy KP, Parent M, Levine MS. The role of dopamine in Huntington's disease. Prog Brain Res [Internet]. 2014;211:235–54. http://www.sciencedirect.com/science/bookseries/00796123, https://doi.org/10.1016/b978-0-444-63425-2.00010-6.

35. Graveland GA, Williams RS, DiFiglia M. Evidence for degenerative and regenerative changes in neostriatal spiny neurons in Huntington's disease. Science. 1985;227(4688):770–3.

36. Ferrante RJ, Kowall NW, Richardson EP Jr. Proliferative and degenerative changes in striatal spiny neurons in Huntington's disease: a combined study using the section-Golgi method and calbindin D28k immunocytochemistry. J Neurosci. 1991;11(12):3877–87.

37. Portera-Cailliau C, Hedreen JC, Price DL, Koliatsos VE. Evidence for apoptotic cell death in Huntington disease and excitotoxic animal models. J Neurosci. 1995;15(5):3775–87.

38. Vonsattel JP, Myers RH, Stevens TJ, Ferrante RJ, Bird ED, Richardson EP Jr. Neuropathological classification of Huntington's disease. J Neuropathol Exp Neurol. 1985;44(6):559–77.

39. De la Monte SM, Vonsattel JP, Richardson EP Jr. Morphometric demonstration of atrophic changes in the cerebral cortex, white matter, and neostriatum in Huntington's disease. J Neuropathol Exp Neurol. 1988;47(5):516–25.

40. Lange H, Thörner G, Hopf A, Schröder KF. Morphometric studies of the neuropathological changes in choreatic diseases. J Neurol Sci. 1976;28(4):401–25.

41. Sapp E, Kegel KB, Aronin N, Hashikawa T, Uchiyama Y, Tohyama K, et al. Early and progressive accumulation of reactive microglia in the Huntington disease brain. J Neuropathol Exp Neurol. 2001;60(2):161–72.

42. Singhrao SK, Neal JW, Morgan BP, Gasque P. Increased complement biosynthesis by microglia and complement activation on neurons in Huntington's disease. Exp Neurol. 1999;159(2):362–76.

43. Gómez-Tortosa E, MacDonald ME, Friend JC, Taylor SA, Weiler LJ, Cupples LA, et al. Quantitative neuropathological changes in presymptomatic Huntington's disease. Ann Neurol. 2001;49(1):29–34.

44. Oyanagi K, Takeda S, Takahashi H, Ohama E, Ikuta F. A quantitative investigation of the substantia nigra in Huntington's disease. Ann Neurol. 1989;26(1):13–9.

45. Heinsen H, Rüb U, Bauer M, Ulmar G, Bethke B, Schüler M, et al. Nerve cell loss in the thalamic mediodorsal nucleus in Huntington's disease. Acta Neuropathol. 1999;97(6):613–22.

46. Kremer HP, Roos RA, Dingjan G, Marani E, Bots GT. Atrophy of the hypothalamic lateral tuberal nucleus in Huntington's disease. J Neuropathol Exp Neurol. 1990;49(4):371–82.

47. Kremer HP, Roos RA, Dingjan GM, Bots GT, Bruyn GW, Hofman MA. The hypothalamic lateral tuberal nucleus and the characteristics of neuronal loss in Huntington's disease. Neurosci Lett. 1991;132(1):101–4.

48. Sotrel A, Paskevich PA, Kiely DK, Bird ED, Williams RS, Myers RH. Morphometric analysis of the prefrontal cortex in Huntington's disease. Neurology. 1991;41(7):1117–23.

49. Hedreen JC, Peyser CE, Folstein SE, Ross CA. Neuronal loss in layers V and VI of cerebral cortex in Huntington's disease. Neurosci Lett. 1991;133(2):257–61.

50. Heinsen H, Strik M, Bauer M, Luther K, Ulmar G, Gangnus D, et al. Cortical and striatal neurone number in Huntington's disease. Acta Neuropathol. 1994;88(4):320–33.

51. Trottier Y, Devys D, Imbert G, Saudou F, An I, Lutz Y, et al. Cellular localization of the Huntington's disease protein and discrimination of the normal and mutated form. Nat Genet. 1995;10(1):104–10.

52. Sharp AH, Loev SJ, Schilling G, Li SH, Li XJ, Bao J, et al. Widespread expression of Huntington's disease gene (IT15) protein product. Neuron. 1995;14(5):1065–74.

53. Gutekunst CA, Levey AI, Heilman CJ, Whaley WL, Yi H, Nash NR, et al. Identification and localization of huntingtin in brain and human lymphoblastoid cell lines with anti-fusion protein antibodies. Proc Natl Acad Sci USA. 1995;92(19):8710–4.

54. DiFiglia M, Sapp E, Chase K, Schwarz C, Meloni A, Young C, et al. Huntingtin is a cytoplasmic protein associated with vesicles in human and rat brain neurons. Neuron. 1995;14(5):1075–81.

55. Gutekunst CA, Li SH, Yi H, Mulroy JS, Kuemmerle S, Jones R, et al. Nuclear and neuropil aggregates in Huntington's disease: relationship to neuropathology. J Neurosci. 1999;19(7):2522–34.

56. Gutekunst CA, Li SH, Yi H, Ferrante RJ, Li XJ, Hersch SM. The cellular and subcellular localization of huntingtin-associated protein 1 (HAP1): comparison with huntingtin in rat and human. J Neurosci. 1998;18(19):7674–86.

57. Kegel KB, Kim M, Sapp E, McIntyre C, Castaño JG, Aronin N, et al. Huntingtin expression stimulates endosomal-lysosomal activity, endosome tubulation, and autophagy. J Neurosci. 2000;20(19):7268–78.

58. Li SH, Li XJ. Aggregation of N-terminal huntingtin is dependent on the length of its glutamine repeats. Hum Mol Genet. 1998;7(5):777–82.

59. Perutz MF, Johnson T, Suzuki M, Finch JT. Glutamine repeats as polar zippers: their possible role in inherited neurodegenerative diseases. Proc Natl Acad Sci USA. 1994;91(12):5355–8.

60. Gourfinkel-An I, Cancel G, Trottier Y, Devys D, Tora L, Lutz Y, et al. Differential distribution of the normal and mutated forms of huntingtin in the human brain. Ann Neurol. 1997;42(5):712–9.

61. Scherzinger E, Lurz R, Turmaine M, Mangiarini L, Hollenbach B, Hasenbank R, et al. Huntingtin-encoded polyglutamine expansions form amyloid-like protein aggregates in vitro and in vivo. Cell. 1997;90(3):549–58.

62. Martindale D, Hackam A, Wieczorek A, Ellerby L, Wellington C, McCutcheon K, et al. Length of huntingtin and its polyglutamine tract influences localization and frequency of intracellular aggregates. Nat Genet. 1998;18(2):150–4.

63. Kuemmerle S, Gutekunst CA, Klein AM, Li XJ, Li SH, Beal MF, et al. Ferrante RJ. Huntington aggregates may not predict neuronal death in Huntington's disease. Ann Neurol. 1999;46(6):842–9.

64. Li SH, Gutekunst CA, Hersch SM, Li XJ. Interaction of huntingtin-associated protein with dynactin P150Glued. J Neurosci. 1998;18(4):1261–9.

65. Li SH, Cheng AL, Zhou H, Lam S, Rao M, Li H, et al. Interaction of Huntington disease protein with transcriptional activator Sp1. Mol Cell Biol. 2002;22(5):1277–87.

66. Li XJ, Li SH, Sharp AH, Nucifora FC Jr, Schilling G, Lanahan A, et al. A huntingtin-associated protein enriched in brain with implications for pathology. Nature. 1995;378(6555):398–402.

67. Li Y, Chin LS, Levey AI, Li L. Huntingtin-associated protein 1 interacts with hepatocyte growth factor-regulated tyrosine kinase substrate and functions in endosomal trafficking. J Biol Chem. 2002;277(31):28212–21.

68. Sittler A, Wälter S, Wedemeyer N, Hasenbank R, Scherzinger E, Eickhoff H, et al. SH3GL3 associates with the Huntingtin exon 1 protein and promotes the formation of polygln-containing protein aggregates. Mol Cell. 1998;2(4):427–36.

69. Kalchman MA, Graham RK, Xia G, Koide HB, Hodgson JG, Graham KC, et al. Huntingtin is ubiquitinated and interacts with a specific ubiquitin-conjugating enzyme. J Biol Chem. 1996;271(32):19385–94.

70. Kalchman MA, Koide HB, McCutcheon K, Graham RK, Nichol K, Nishiyama K, et al. HIP1, a human homologue of S. cerevisiae Sla2p, interacts with membrane-associated huntingtin in the brain. Nat Genet. 1997;16(1):44–53.

71. Wanker EE, Rovira C, Scherzinger E, Hasenbank R, Wälter S, Tait D, et al. HIP-I: a huntingtin interacting protein isolated by the yeast two-hybrid system. Hum Mol Genet. 1997;6(3):487–95.

72. Metzler M, Legendre-Guillemin V, Gan L, Chopra V, Kwok A, McPherson PS, et al. HIP1 functions in clathrin-mediated endocytosis through binding to clathrin and adaptor protein 2. J Biol Chem. 2001;276(42):39271–6.

73. Lunkes A, Lindenberg KS, Ben-Haïem L, Weber C, Devys D, Landwehrmeyer GB, et al. Proteases acting on mutant huntingtin generate cleaved products that differentially build up cytoplasmic and nuclear inclusions. Mol Cell. 2002;10(2):259–69.

74. Zucker B, Kama JA, Kuhn A, Thu D, Orlando LR, Dunah AW, et al. Decreased Lin7b expression in layer 5 pyramidal neurons may contribute to impaired corticostriatal connectivity in huntington disease. J Neuropathol Exp Neurol [Internet]. 2010;69(9):880–95. https://academic.oup.com/jnen, https://doi.org/10.1097/nen.0b013e3181ed7a41.

75. Chakraborty J, Rajamma U, Mohanakumar KP. A mitochondrial basis for Huntington's disease: therapeutic prospects. Mol Cell Biochem [Internet]. 2014;389(1–2):277–91. http://link.springer.com/journal/11010 https://doi.org/10.1007/s11010-013-1951-9.

76. Gladding CM. Raymond LA Mechanisms underlying NMDA receptor synaptic/extrasynaptic distribution and function. Mol Cell Neurosci. 2011;48(4):308–20.

77. Deyts C, Galan-Rodriguez B, Martin E, Bouveyron N, Roze E, Charvin D, et al. Dopamine D2 receptor stimulation potentiates PolyQ-Huntingtin-induced mouse striatal neuron dysfunctions via Rho/ROCK-II activation. PLoS One [Internet]. 2009;4(12):e8287. http://www.ncbi.nlm.nih.gov/pmc/journals/440/.

78. Strand AD, Baquet ZC, Aragaki AK, Holmans P, Yang L, Cleren C, et al. Expression profiling of Huntington's disease models suggests that brain-derived neurotrophic factor depletion plays a major role in striatal degeneration. J Neurosci [Internet]. 2007;27(43):11758–68. http://www.jneurosci.org/, https://doi.org/10.1523/jneurosci.2461-07.2007.

79. Li JY, Conforti L. Axonopathy in Huntington's disease. Exp Neurol [Internet]. 2013;246:62–71. http://www.sciencedirect.com/science/journal/00144886, https://doi.org/10.1016/j.expneurol.2012.08.010.

80. Crotti A, Glass CK. The choreography of neuroinflammation in Huntington's disease. Trends Immunol. 2015;36(6):364–73.

81. Cortes CJ, La Spada AR. The many faces of autophagy dysfunction in Huntington's disease: from mechanism to therapy. Drug Discov Today [Internet]. 2014;19(7):963–71. http://sciencedirect.com/science/journal/13596446, https://doi.org/10.1016/j.drudis.2014.02.014.

82. Ellrichmann G, Reick C, Saft C, Linker RA. The role of the immune system in Huntington's disease. Clin Dev Immunol [Internet]. 2013;2013:541259. http://www.ncbi.nlm.nih.gov/pmc/journals/499/#cdi, https://doi.org/10.1155/2013/541259.

83. Rosas HD, Liu AK, Hersch S, Glessner M, Ferrante RJ, Salat DH, et al. Regional and progressive thinning of the cortical ribbon in Huntington's disease. Neurology. 2002;58(5):695–701.

84. Rosas HD, Koroshetz WJ, Chen YI, Skeuse C, Vangel M, Cudkowicz ME, et al. Evidence for more widespread cerebral pathology in early HD: an MRI-based morphometric analysis. Neurology. 2003;60(10):1615–20.

85. Thieben MJ, Duggins AJ, Good CD, Gomes L, Mahant N, Richards F, et al. The distribution of structural neuropathology in pre-clinical Huntington's disease. Brain. 2002;125(8):1815–28.

86. Klöppel S, Henley SM, Hobbs NZ, Wolf RC, Kassubek J, Tabrizi SJ, et al. Magnetic resonance imaging of Huntington's disease: preparing for clinical trials. Neuroscience [Internet]. 2009;164(1):205–19. http://sciencedirect.com/science/journal/03064522, https://doi.org/10.1016/j.neuroscience.2009.01.045.

87. Pagano G, Niccolini F, Politis M. Current status of PET imaging in Huntington's disease. Eur J Nucl Med Mol Imaging [Internet]. 2016;43(6):1171–82. http://link.springer.com/journal/259, https://doi.org/10.1007/s00259-016-3324-6.

88. Nopoulos PC. Huntington disease: a single-gene degenerative disorder of the striatum. Dialogues Clin Neurosci. 2016;18(1):91–8.

89. Mahant N, McCusker EA, Byth K, Graham S, Huntington Study Group. Huntington's disease: clinical correlates of disability and progression. Neurology. 2003;61(8):1085–92.

90. Novak MJ, Tabrizi SJ. Huntington's disease. BMJ [Internet]. 2010;340:c3109. http://www.ncbi.nlm.nih.gov/pmc/journals/3/ https://doi.org/10.1136/bmj.c3109.

91. Quarrell OW, Nance MA, Nopoulos P, Paulsen JS, Smith JA, Squitieri F. Managing juvenile Huntington's disease. Neurodegener Dis Manag [Internet]. 2013;3(3). http://www.future-medicine.com/loi/nmt, https://doi.org/10.2217/nmt.13.18.

92. Cui SS, Ren RJ, Wang Y, Wang G, Chen SD. Tics as an initial manifestation of juvenile Huntington's disease: case report and literature review. BMC Neurol [Internet]. 2017;17(1):152. http://www.ncbi.nlm.nih.gov/pmc/journals/48/, https://doi.org/10.1186/s12883-017-0923-1.

93. Kirkwood SC, Siemers E, Stout JC, Hodes ME, Conneally PM, Christian JC, et al. Longitudinal cognitive and motor changes among presymptomatic Huntington disease gene carriers. Arch Neurol. 1999;56(5):563–8.

94. Khedraki A, Reed EJ, Romer SH, Wang Q, Romine W, Rich MM, Talmadge RJ, Voss AA. Depressed synaptic transmission and reduced vesicle release sites in Huntington's

disease neuromuscular junctions. J Neurosci. 2017;37(34):8077–91. https://doi.org/10.1523/JNEUROSCI.0313-17.2017.

95. Danoudis M, Iansek R. Gait in Huntington's disease and the stride length-cadence relationship. BMC Neurol [Internet]. 2014;14:161. http://www.ncbi.nlm.nih.gov/pmc/journals/48/, https://doi.org/10.1186/s12883-014-0161-8.

96. Simpson JA, Lovecky D, Kogan J, Vetter LA, Yohrling GJ. Survey of the Huntington's disease patient and caregiver community reveals most impactful symptoms and treatment needs. J Huntingtons Dis [Internet]. 2016;5(4):395–403. http://content.iospress.com/journals/journal-of-huntingtons-disease.

97. Lange KW, Sahakian BJ, Quinn NP, Marsden CD, Robbins TW. Comparison of executive and visuospatial memory function in Huntington's disease and dementia of Alzheimer type matched for degree of dementia. J Neurol Neurosurg Psychiatry. 1995;58(5):598–606.

98. Paulsen JS, Butters N, Sadek JR, Johnson SA, Salmon DP, Swerdlow NR, et al. Distinct cognitive profiles of cortical and subcortical dementia in advanced illness. Neurology. 1995;45(5):951–6.

99. Bamford KA, Caine ED, Kido DK, Cox C, Shoulson I. A prospective evaluation of cognitive decline in early Huntington's disease: functional and radiographic correlates. Neurology. 1995;45(10):1867–73.

100. Paulsen JS. Cognitive impairment in Huntington disease: diagnosis and treatment. Curr Neurol Neurosci Rep [Internet]. 2011;11(5):474–83. http://link.springer.com/journal/11910, https://doi.org/10.1007/s11910-011-0215-x.

101. Schneider SA, Wilkinson L, Bhatia KP, Henley SM, Rothwell JC, Tabrizi SJ. et al. Abnormal explicit but normal implicit sequence learning in premanifest and early Huntington's disease. Mov Disord [Internet]. 2010;25(10):1343–9. http://onlinelibrary.wiley.com/journal/10.1002/(ISSN)1531-8257, https://doi.org/10.1002/mds.22692.

102. Huntington Study Group [No authors listed]. Unified Huntington's disease rating scale: reliability and consistency. Mov Disord. 1996;11(2):136–42.

103. Paulsen JS, Langbehn DR, Stout JC, Aylward E, Ross CA, Nance M, et al. Detection of Huntington's disease decades before diagnosis: the Predict-HD study. J Neurol Neurosurg Psychiatry. 2008;79(8):874–80.

104. Foroud T, Siemers E, Kleindorfer D, Bill DJ, Hodes ME, Norton JA, et al. Cognitive scores in carriers of Huntington's disease gene compared to noncarriers. Ann Neurol. 1995;37(5):657–64.

105. Unmack Larsen I, Vinther-Jensen T, Gade A, Nielsen JE, Vogel A. Assessing impairment of executive function and psychomotor speed in premanifest and manifest Huntington's disease gene-expansion carriers. J Int Neuropsychol Soc [Internet]. 2015;21(3):193–202. https://www.cambridge.org/core/journals/journal-of-the-international-neuropsychological-society, https://doi.org/10.1017/s1355617715000090.

106. Duff K, Paulsen JS, Beglinger LJ, Langbehn DR, Wang C, Stout JC, et al. "Frontal" behaviors before the diagnosis of Huntington's disease and their relationship to markers of disease progression: evidence of early lack of awareness. J Neuropsychiatry Clin Neurosci [Internet]. 2010;22(2):196–207. http://neuro.psychiatryonline.org/journal.aspx?journalid=62, https://doi.org/10.1176/appi.neuropsych.22.2.196.

107. Mörkl S, Müller NJ, Blesl C, Wilkinson L, Tmava A, Wurm W, et al. Problem solving, impulse control and planning in patients with early- and late-stage Huntington's disease. Eur Arch Psychiatry Clin Neurosci [Internet]. 2016;266(7):663–71. http://link.springer.com/journal/406, https://doi.org/10.1007/s00406-016-0707-4.

108. Dumas EM, van den Bogaard SJ, Middelkoop HA, Roos RA. A review of cognition in Huntington's disease. Front Biosci. 2013;1(5):1–18.

109. Nance M, Paulsen J, Rosenblatt A, Wheelock V. A physician's guide to the management of Huntington's disease. New York: Huntigton's Disease Society of America; 2012. p. 55.

110. Feigin A, Ghilardi MF, Huang C, Ma Y, Carbon M, Guttman M, Paulsen JS, Ghez CP, Eidelberg D. Preclinical Huntington's disease: compensatory brain responses during learning. Ann Neurol. 2006;59(1):53–9.

111. Johnson SA, Stout JC, Solomon AC, Langbehn DR, Aylward EH, Cruce CB, et al. Beyond disgust: impaired recognition of negative emotions prior to diagnosis in Huntington's disease. Brain. 2007;130(Pt 7):1732–44.

112. Brandt J, Shpritz B, Munro CA, Marsh L, Rosenblatt A. Differential impairment of spatial location memory in Huntington's disease. J Neurol Neurosurg Psychiatry. 2005;76(11):1516–9.

113. Hamilton JM, Murphy C, Paulsen JS. Odor detection, learning, and memory in Huntington's disease. J Int Neuropsychol Soc. 1999;5(7):609–15.

114. Rowe KC, Paulsen JS, Langbehn DR, Duff K, Beglinger LJ, Wang C, et al. Self-paced timing detects and tracks change in prodromal Huntington disease. Neuropsychology [Internet]. 2010;24(4):435–42. http://search.ebscohost.com/direct.asp?db=pdh&jid=NEU-&scope=site, https://doi.org/10.1037/a0018905.

115. Sitek EJ, Sołtan W, Wieczorek D, Schinwelski M, Robowski P, Reilmann R, et al. Self-awareness of motor dysfunction in patients with Huntington's disease in comparison to Parkinson's disease and cervical dystonia. J Int Neuropsychol Soc [Internet]. 2011;17(5):788–95. https://www.cambridge.org/core/journals/journal-of-the-international-al-neuropsychological-society, https://doi.org/10.1017/s1355617711000725.

116. Frank E, Morrow-Odom KL, Abramson RK, Cuturic M. Central auditory and visual processing in Huntington's disease. J Med Speech Lang Pathol [Internet]. 2009;17(1):3. https://www.pluralpublishing.com/journals_JMSLP.htm.

117. Saldert C, Fors A, Ströberg S, Hartelius L. Comprehension of complex discourse in different stages of Huntington's disease. Int J Lang Commun Disord [Internet]. 2010;45(6):656–69. http://onlinelibrary.wiley.com/journal/10.1111/(ISSN)1460-6984, https://doi.org/10.3109/13682820903494742.

118. Shiwach R. Psychopathology in Huntington's disease patients. Acta Psychiatr Scand. 1994;90(4):241–6.

119. Van Duijn E, Craufurd D, Hubers AA, Giltay EJ, Bonelli R, Rickards H, et al. Neuropsychiatric symptoms in a European Huntington's disease cohort (REGISTRY). J Neurol Neurosurg Psychiatry [Internet]. 2014;85(12):1411–8. http://www.ncbi.nlm.nih.gov/pmc/journals/192/, https://doi.org/10.1136/jnnp-2013-307343.

120. Ghosh R, Tabrizi SJ. Clinical aspects of Huntington's disease. Curr Top Behav Neurosci. 2015;22:3–31.

121. Vaccarino AL, Sills T, Anderson KE, Bachoud-Lévi AC, Borowsky B, Craufurd D, et al. Assessment of depression, anxiety and apathy in prodromal and early huntington disease. PLoS Curr [Internet]. 2011;3:RRN1242. http://www.ncbi.nlm.nih.gov/pmc/?term=%22PLoS+Curr%22%5Bjournal%5D.

122. Epping EA, Kim JI, Craufurd D, Brashers-Krug TM, Anderson KE, McCusker E. Longitudinal psychiatric symptoms in prodromal Huntington's disease: a decade of data. Am J Psychiatry [Internet]. 2016;173(2):184–92. http://ajp.psychiatryonline.org/journal.aspx?journalid=13, https://doi.org/10.1176/appi.ajp.2015.14121551.

123. Paulsen JS, Nehl C, Hoth KF, Kanz JE, Benjamin M, Conybeare R, et al. Depression and stages of Huntington's disease. J Neuropsychiatry Clin Neurosci. 2005;17(4):496–502.

124. Mendez MF. Huntington's disease: update and review of neuropsychiatric aspects. Int J Psychiatry Med. 1994;24(3):189–208.

125. Fisher CA, Sewell K, Brown A, Churchyard A. Aggression in Huntington's disease: a systematic review of rates of aggression and treatment methods. J Huntingtons Dis [Internet]. 2014;3(4):319–32. http://content.iospress.com/journals/journal-of-huntingtons-disease, https://doi.org/10.3233/jhd-140127.

126. Lipe H, Schultz A, Bird TD. Risk factors for suicide in Huntingtons disease: a retrospective case controlled study. Am J Med Genet. 1993;48(4):231–3.
127. Almqvist EW, Bloch M, Brinkman R, Craufurd D, Hayden MR. A worldwide assessment of the frequency of suicide, suicide attempts, or psychiatric hospitalization after predictive testing for Huntington disease. Am J Hum Genet. 1999;64(5):1293–304.
128. Wetzel HH, Gehl CR, Dellefave-Castillo L, Schiffman JF, Shannon KM, Paulsen JS, et al. Suicidal ideation in Huntington disease: the role of comorbidity. Psychiatry Res [Internet]. 2011;188(3):372–6. http://www.sciencedirect.com/science/journal/09254927, https://doi.org/10.1016/j.psychres.2011.05.006.
129. Paulsen JS, Hoth KF, Nehl C, Stierman L. Critical periods of suicide risk in Huntington's disease. Am J Psychiatry. 2005;162(4):725–31.
130. Farrer LA. Suicide and attempted suicide in Huntington disease: implications for preclinical testing of persons at risk. Am J Med Genet. 1986;24(2):305–11.
131. Cummings J. Behavioral and psychiatric symptoms associated with Huntington disease. In: Weiner WJ, Lang AE, editors. Behavioral neurology of movement disorders. New York: Raven Press; 1995. p. 179–86.
132. Schoenfeld M, Myers RH, Cupples LA, Berkman B, Sax DS, Clark E. Increased rate of suicide among patients with Huntington's disease. J Neurol Neurosurg Psychiatry. 1984;47(12):1283–7.
133. Lovestone S, Hodgson S, Sham P, Differ AM, Levy R. Familial psychiatric presentation of Huntington's disease. J Med Genet. 1996;33(2):128–31.
134. Nagel M, Rumpf HJ, Kasten M. Acute psychosis in a verified Huntington disease gene carrier with subtle motor signs: psychiatric criteria should be considered for the diagnosis. Gen Hosp Psychiatry [Internet]. 2014;36(3):361.e3–4. http://www.sciencedirect.com/science/journal/01638343, https://doi.org/10.1016/j.genhosppsych.2014.01.008.
135. Cummings JL, Cunningham K. Obsessive-compulsive disorder in Huntington's disease. Biol Psychiatry. 1992;31(3):263–70.
136. Cuturic M, Abramson RK, Vallini D, Frank EM, Shamsnia M. Sleep patterns in patients with Huntington's disease and their unaffected first-degree relatives: a brief report. Behav Sleep Med [Internet]. 2009;7(4):245–54. http://www.tandfonline.com/loi/hbsm, https://doi.org/10.1080/15402000903190215.
137. Wiegand M, Möller AA, Lauer CJ, Stolz S, Schreiber W, Dose M, et al. Nocturnal sleep in Huntington's disease. J Neurol. 1991;238(4):203–8.
138. Pflanz S, Besson JA, Ebmeier KP, Simpson S. The clinical manifestation of mental disorder in Huntington's disease: a retrospective case record study of disease progression. Acta Psychiatr Scand. 1991;83(1):53–60.
139. Kirkwood SC, Siemers E, Viken R, Hodes ME, Conneally PM, Christian JC, et al. Longitudinal personality changes among presymptomatic Huntington disease gene carriers. Neuropsychiatry Neuropsychol Behav Neurol. 2002;15(3):192–7.
140. Cardoso F, Seppi K, Mair KJ, Wenning GK, Poewe W. Seminar on choreas. Lancet Neurol. 2006;5(7):589–602.
141. Rosencrantz R, Schilsky M. Wilson disease: pathogenesis and clinical considerations in diagnosis and treatment. Semin Liver Dis [Internet]. 2011;31(3):245–59. https://www.thieme-connect.de/products/ejournals/journal/10.1055/s-00000069, https://doi.org/10.1055/s-0031-1286056.
142. Walker RH, Jung HH, Danek A. Neuroacanthocytosis. Handb Clin Neurol. 2011;100:141–51.
143. Stevanin G, Brice A. Spinocerebellar ataxia 17 (SCA17) and Huntington's disease-like 4 (HDL4). Cerebellum [Intertnet]. 2008;7(2):170–8. http://link.springer.com/journal/12311, https://doi.org/10.1007/s12311-008-0016-1.
144. Iizuka R, Hirayama K, Maehara KA. Dentato-rubro-pallido-luysian atrophy: a clinico-pathological study. Huntingtin is ubiquitinated and interacts with a specific ubiquitin-conjugating enzyme. J Neurol Neurosurg Psychiatry. 1984;47(12):1288–98.

145. Hardie RJ, Pullon HW, Harding AE, Owen JS, Pires M, Daniels GL, et al. Neuroacanthocytosis. A clinical, haematological and pathological study of 19 cases. Brain. 1991;114(Pt 1A):13–49.

146. Laplanche JL, Hachimi KH, Durieux I, Thuillet P, Defebvre L, Delasnerie-Laupretre N, et al. Prominent psychiatric features and early onset in an inherited prion disease with a new insertional mutation in the prion protein gene. Brain. 1999;122:2375–86.

147. Moore RC, Xiang F, Monaghan J, Han D, Zhang Z, Edstrom L, et al. Huntington disease phenocopy is a familial prion disease. Am J Hum Genet. 2001;69:1385–8.

148. Margolis RL, Rudnicki DD, Holmes SE. Huntington's disease like-2: review and update. Acta Neurol Taiwan. 2005;14:1–8.

149. Kremer B, Goldberg P, Andrew SE, Theilmann J, Telenius H, Zeisler J, et al. A worldwide study of the Huntington's disease mutation. The sensitivity and specificity of measuring CAG repeats. N Engl J Med. 1994;330(20):1401–6.

150. Terrenoire G. Huntington's disease and the ethics of genetic prediction. J Med Ethics. 1992;18(2):79–85.

151. Hayden MR, Bloch M, Wiggins S. Psychological effects of predictive testing for Huntington's disease. Adv Neurol. 1995;65:201–10.

152. Harper PS. Clinical consequences of isolating the gene for Huntington's disease. BMJ. 1993;307(6901):397–8.

153. International Huntington Association (IHA) and the World Federation of Neurology (WFN) Research Group on Huntington's Chorea. [No authors listed]. Guidelines for the molecular genetics predictive test in Huntington's disease. Neurology. 1994;44(8):1533–6.

154. Nance M, Paulsen J, Rosenblatt A, Wheelock V. A physician's guide to the management of Huntington's disease. New York: Huntigton's Disease Society of America; 2012. p. 16–23.

155. Richards FH. Maturity of judgement in decision making for predictive testing for nontreatable adult-onset neurogenetic conditions: a case against predictive testing of minors. Clin Genet. 2006;70(5):396–401.

156. Oster E, Dorsey ER, Bausch J, Shinaman A, Kayson E, Oakes D, et al. Fear of health insurance loss among individuals at risk for Huntington disease. Am J Med Genet A [Internet]. 2008;146A(16):2070–7. http://onlinelibrary.wiley.com/journal/10.1002/(ISSN)1552-4833, https://doi.org/10.1002/ajmg.a.32422.

157. Quaid KA, Eberly SW, Kayson-Rubin E, Oakes D, Shoulson I; Huntington Study Group PHAROS Investigators and Coordinators. Factors related to genetic testing in adults at risk for Huntington disease: the prospective Huntington at-risk observational study (PHAROS). Clin Genet [Internet]. 2017;91(6):824–31. http://onlinelibrary.wiley.com/journal/10.1111/(ISSN)1399-0004, https://doi.org/10.1111/cge.12893.

158. De Die-Smulders CE, de Wert GM, Liebaers I, Tibben A, Evers-Kiebooms G. Reproductive options for prospective parents in families with Huntington's disease: clinical, psychological and ethical reflections. Hum Reprod Update [Internet]. 2013;19(3):304–15. https://academic.oup.com/humupd, https://doi.org/10.1093/humupd/dms058.

159. Bustamante-Aragonés A, Rodríguez de Alba M, Perlado S, Trujillo-Tiebas MJ, Arranz JP, Díaz-Recasens et al. Non-invasive prenatal diagnosis of single-gene disorders from maternal blood. Gene [Internet]. 2012;504(1):144–9. http://sciencedirect.com/science/journal/03781119, https://doi.org/10.1016/j.gene.2012.04.045.

160. Van den Oever JM, Bijlsma EK, Feenstra I, Muntjewerff N, Mathijssen IB, Bakker E, et al. Noninvasive prenatal diagnosis of Huntington disease: detection of the paternally inherited expanded CAG repeat in maternal plasma. Prenat Diagn [Internet]. 2015;35(10):945–9. http://onlinelibrary.wiley.com/journal/10.1002/(ISSN)1097-0223, https://doi.org/10.1002/pd.4593.

161. Schultz JL, Kamholz JA, Moser DJ, Feely SM, Paulsen JS, Nopoulos PC. Substance abuse may hasten motor onset of Huntington disease: evaluating the enroll-HD database. Neurology [Internet]. 2017;88(9):909–15. http://ovidsp.ovid.com/ovidweb.cgi?T=JS&NEWS=n&CSC=Y&PAGE=toc&D=yrovft&AN=00006114-000000000-00000, https://doi.org/10.1212/wnl.0000000000003661.

162. Frese S, Petersen JA, Ligon-Auer M, Mueller SM, Mihaylova V, Gehrig SM, et al. Exercise effects in Huntington disease. J Neurol [Internet]. 2017;264(1):32–39. http://link.springer. com/journal/415, https://doi.org/10.1007/s00415-016-8310-1.

163. Rivadeneyra J, Cubo E, Gil C, Calvo S, Mariscal N, Martínez A. Factors associated with Mediterranean diet adherence in Huntington's disease. Clin Nutr ESPEN [Internet]. 2016;12:e7–13. http://www.sciencedirect.com/science/journal/24054577, https://doi. org/10.1016/j.clnesp.2016.01.001.

164. Shannon KM, Fraint A. Therapeutic advances in Huntington's disease. Mov Disord [Internet]. 2015;30(11):1539–46. http://onlinelibrary.wiley.com/journal/10.1002/(ISSN)1531-8257, https://doi.org/10.1002/mds.26331.

165. Marosz A, Chlubek D. The risk of abuse of vitamin supplements. Ann Acad Med Stetin. 2014;60(1):60–4.

166. Klimek ML, Rohs G, Young L, Suchowersky O, Trew M. Multidisciplinary approach to management of a hereditary neurodegenerative disorder: Huntington disease. Axone. 1997;19(2):34–8.

167. Quinn L, Busse M, Carrier J, Fritz N, Harden J, Hartel L, et al. Physical therapy and exercise interventions in Huntington's disease: a mixed methods systematic review protocol. JBI Database System Rev Implement Rep [Internet]. 2017;15(7):1783–99. http://ovidsp.ovid.com/ovidweb. cgi?T=JS&CSC=Y&NEWS=N&PAGE=toc&SEARCH=01938924-201606000-00000. kc&LINKTYPE=asBody&LINKPOS=1&D=yrovft, https://doi.org/10.11124/jbisrir-2016-003274.

168. Piira A, van Walsem MR, Mikalsen G, Øie L, Frich JC, Knutsen S. Effects of a two-year intensive multidisciplinary rehabilitation program for patients with Huntington's disease: a prospective intervention study. Version 2. PLoS Curr [Internet]. 2014 [revised 2014 Jan 1];6. pii: ecurrents.hd.2c56ceef7f9f8e239a59ecf2d94cddac. http://www.ncbi.nlm.nih. gov/pmc/?term=%22PLoS+Curr%22%5Bjournal%5D, https://doi.org/10.1371/currents. hd.2c56ceef7f9f8e239a59ecf2d94cddac.

169. Zinzi P, Salmaso D, De Grandis R, Graziani G, Maceroni S, Bentivoglio A, et al. Effects of an intensive rehabilitation programme on patients with Huntington's disease: a pilot study. Clin Rehabil. 2007;21(7):603–13.

170. Cruickshank TM, Thompson JA, Domínguez D JF, Reyes AP, Bynevelt M, Georgiou-Karistianis N, et al. The effect of multidisciplinary rehabilitation on brain structure and cognition in Huntington's disease: an exploratory study. Brain Behav [Internet]. 2015;5(2):e00312. http://www.ncbi.nlm.nih.gov/pmc/journals/1650/, https://doi. org/10.1002/brb3.312.

171. Downing NR, Goodnight S, Chae S, Perlmutter JS, McCormack M, Hahn E, et al. Factors associated with end-of-life planning in Huntington disease. Am J Hosp Palliat Care [Internet]. 2017:1049909117708195. http://journals.sagepub.com/home/ajh, https://doi. org/10.1177/1049909117708195.

172. Dellefield ME, Ferrini R. Promoting Excellence in end-of-life care: lessons learned from a cohort of nursing home residents with advanced Huntington disease. J Neurosci Nurs [Internet]. 2011;43(4):186–92. http://ovidsp.ovid.com/ovidweb.cgi?T= JS&NEWS=n&CSC=Y&PAGE=toc&D=yrovft&AN=01376517-000000000-00000, https://doi.org/10.1097/jnn.0b013e3182212a52.

173. Veenhuizen RB, Kootstra B, Vink W, Posthumus J, van Bekkum P, Zijlstra M, et al. Coordinated multidisciplinary care for ambulatory Huntington's disease patients. Evaluation of 18 months of implementation. Orphanet J Rare Dis [Internet]. 2011;6:77. http://www. ncbi.nlm.nih.gov/pmc/journals/401/, https://doi.org/10.1186/1750-1172-6-77.

174. Bonelli RM, Wenning GK. Pharmacological management of Huntington's disease: an evidence-based review. Curr Pharm Des. 2006;12(21):2701–20.

175. Mestre T, Ferreira J, Coelho MM, Rosa M, Sampaio C. Therapeutic interventions for symptomatic treatment in Huntington's disease. Cochrane Database Syst Rev [Internet]. 2009;(3):CD006456. http://www.thecochranelibrary.com/view/0/index.html, https://doi.org/10.1002/14651858.cd006456.pub2.

176. Scheifer J, Werner CJ, Reetz K. Clinical diagnosis and management in early Huntington's disease: a review. Degener Neurol Neuromuscul Dis. 2015;5: 37–50.

177. Cuturic M, Abramson RK, Moran RR, Hardin JW, Frank EM, Sellers AA. Serum carnitine levels and levocarnitine supplementation in institutionalized Huntington's disease patients. Neurol Sci. 2013;34(1):93–8.

178. Nance M, Paulsen J, Rosenblatt A, Wheelock V. A physician's guide to the management of Huntington's disease. New York: Huntigton's Disease Society of America; 2012. p. 64–58.

179. Squitieri F, Cannella M, Porcellini A, Brusa L, Simonelli M, Ruggieri S. Short-term effects of olanzapine in Huntington disease. Neuropsychiatry Neuropsychol Behav Neurol. 2001;14(1):69–72.

180. Duff K, Beglinger LJ, O'Rourke ME, Nopoulos P, Paulson HL, Paulsen JS. Risperidone and the treatment of psychiatric, motor, and cognitive symptoms in Huntington's disease. Ann Clin Psychiatry [Internet]. 2008;20(1):1–3. http://portico.org/stable?cs=ISSN_10401237, https://doi.org/10.1080/10401230701844802.

181. Huntington Study Group. Tetrabenazine as antichorea therapy in Huntington disease: a randomized controlled trial. Neurology. 2006;66(3):366–72.

182. Huntington Study Group, Frank S, Testa CM, Stamler D, Kayson E, Davis C, et al. Effect of deutetrabenazine on chorea among patients with Huntington disease: a randomized clinical trial. JAMA. 2016;316(1):40–50. http://www.ncbi.nlm.nih.gov/pmc/journals/48/, https://doi.org/10.1001/jama.2016.8655.

183. Claassen DO, Carroll B, De Boer LM, Wu E, Ayyagari R, Gandhi S, et al. Indirect tolerability comparison of Deutetrabenazine and Tetrabenazine for Huntington disease. J Clin Mov Disord [Inernet]. 2017;4:3. http://www.clinicalmovementdisorders.com/, https://doi.org/10.1186/s40734-017-0051-5.

184. Nance M, Paulsen J, Rosenblatt A, Wheelock V. A physician's guide to the management of Huntington's disease. New York: Huntigton's Disease Society of America; 2012. p. 42–4.

185. Travessa AM, Rodrigues FB, Mestre TA, Ferreira JJ. Fifteen years of clinical trials in Huntington's disease: a very low clinical drug development success rate. J Huntingtons Dis [Internet]. 2017;6(2):157–63. http://content.iospress.com/journals/journal-of-huntingtons-disease, https://doi.org/10.3233/jhd-170245.

186. Paulsen JS, Long JD, Johnson HJ, Aylward EH, Ross CA, Williams JK, et al. Clinical and biomarker changes in premanifest Huntington disease show trial feasibility: a decade of the PREDICT-HD Study. Front Aging Neurosci [Internet]. 2014;6:78. http://www.ncbi.nlm.nih.gov/pmc/journals/1239/, https://doi.org/10.3389/fnagi.2014.00078.

187. Byrne LM, Rodrigues FB, Blennow K, Durr A, Leavitt BR, Roos RAC, et al. Neurofilament light protein in blood as a potential biomarker of neurodegeneration in Huntington's disease: a retrospective cohort analysis. Lancet Neurol [Internet]. 2017;16(8):601–9. http://www.sciencedirect.com/science/journal/14744422, https://doi.org/10.1016/s1474-4422(17)30124-2.

188. Rodrigues FB, Byrne L, McColgan P, Robertson N, Tabrizi SJ, Leavitt BR, et al. Cerebrospinal fluid total tau concentration predicts clinical phenotype in Huntington's disease. J Neurochem [Internet]. 2016;139(1):22–5. http://onlinelibrary.wiley.com/journal/10.1111/(ISSN)1471-4159, https://doi.org/10.1111/jnc.13719.

189. Wild EJ, Boggio R, Langbehn D, Robertson N, Haider S, Miller JR, et al. Quantification of mutant huntingtin protein in cerebrospinal fluid from Huntington's disease patients. J Clin Invest [Internet]. 2015;125(5):1979–86. http://www.ncbi.nlm.nih.gov/pmc/journals/120/, https://doi.org/10.1172/jci80743.

190. Holmans P, Stone T. Using genomic data to find disease-modifying loci in Huntington's disease (HD). Methods Mol Biol [Internet]. 2018;1780:443–61. https://link.springer.com/bookseries/7651, https://doi.org/10.1007/978-1-4939-7825-0_20.

191. Correia K, Harold D, Kim KH, Holmans P, Jones L, Orth M, et al. The genetic modifiers of motor onset age (GeM MOA) website: genome-wide association analysis for genetic modifiers of Huntington's disease. J Huntingtons Dis [Internet]. 2015;4(3):279–84. http://content.iospress.com/journals/journal-of-huntingtons-disease, https://doi.org/10.3233/jhd-150169.

192. Keiser MS, Kordasiewicz HB, McBride JL. Gene suppression strategies for dominantly inherited neurodegenerative diseases: lessons from Huntington's disease and spinocerebellar ataxia. Hum Mol Genet [Internet]. 2016 Apr 15;25(R1):R53–64. https://academic.oup.com/hmg, https://doi.org/10.1093/hmg/ddv442. Epub 2015 Oct 26.

193. Tabrizi SJ, Leavitt BR, Landwehrmeyer GB, Wild EJ, Saft C, Barker RA, Blair NF et al. Targeting Huntington expression in patients with Huntington's disease. N Engl J Med [Internet]. 2019;380(24):2307–16. http://www.nejm.org/, https://doi.org/10.1056/nejmoa1900907. Epub 2019 May 6.

Non-motor Symptoms in Parkinson's Disease

Vladimira Vuletić

Introduction

Parkinson's disease (PD) is a chronic neurodegenerative disease characterized by motor symptoms (bradykinesia, rigidity, rest tremor, postural instability) and non-motor symptoms (pain, autonomic dysfunction, cognitive function, depression, fatigue, apathy, sleep disturbances). Although we usually think about motor symptoms (MS) when treating Parkinson's disease (PD), the non-motor symptoms (NMS) may precede the appearance of motor symptoms by several years and are a major cause of disability for PD patients [1]. But they are often unrecognized and untreated, although they can be successfully treated by medications and invasive methods, like deep brain stimulation (DBS) [2, 3]. Nevertheless, NMS correlate with advancing age and disease severity [4]. A modern holistic approach to treat PD should include the recognition and assessment of NMS. Probably, non-motor symptoms and their management will become more important as the average life expectancy of the population increases especially considering their impact also on the quality of life, institutionalization rates, health economics and mortality rates [5, 6]. We have new scales that have been developed which allow us the accurate qualitative and quantitative measurement of all NMS in PD patients (NMS Questionnaire-NMSQ and NMS scale—NMSS) [7, 8] and there are a lot of scales for each NMS [9, 10]. PD results from degeneration of the substantia nigra pars compacta and the consequent dysfunction of the dopaminergic nigrostriatal pathway and presence of Lewy bodies (misfolded α-synuclein), but additional involvement of serotonergic, noradrenergic, and cholinergic pathways are important especially in non-motor symptoms [11].

V. Vuletić (✉)
Clinical Department of Neurology, University Hospital Centre Rijeka, Rijeka, Croatia
e-mail: vladimira.vuletic@gmail.com

Medical Faculty University of Rijeka, Rijeka, Croatia

Cognitive Problems and Dementia

Cognitive problems from mild dysfunction (mild cognitive impairment) to dementia (Parkinson's disease dementia- PDD) are among the most important NMS and can be present in up to 83% of PD patients [12]. The point prevalence of dementia is around 30% and the incidence and risk of dementia in PD is 4–6 times higher than in the age-matched controls [13]. The risks for developing dementia are older age of onset, akinetic-rigid subtype, motor impairment severity, men gender, gait problems, sleep behavior disorder, cardiovascular autonomic [13–19]. The most common cognitive decline in people with PD is in executive, attentional, visuospatial domains, and memory [20]. Dementia in PD is associated with institutionalization, quality of life, caregiver burden, and health-related costs [20]. The International Parkinson and Movement Disorder Society (MDS) published a review and clinical criteria for PD dementia (PDD) and diagnostic criteria for mild cognitive impairment in PD (PD-MCI in 2007 [13, 20]), that can help us in accurately detecting cognitive problems in PD patients. Most neuropathological studies indicate Braak's hypothesis of spreading α-synuclein pathology (Lewy bodies) from sites in the lower brainstem or even extracranially from the gut or other areas innervated by the vagus nucleus to the midbrain, forebrain, limbic structures, and neocortical regions [22, 23] but Alzheimer's disease pathology is also often seen in up to one-third of patients with PD and can contribute to dementia [24, 25]. Nevertheless, considering neurotransmitter changes in PDD beside severe dopamine deficits, there is a marked loss of limbic and cortically projecting dopamine, noradrenaline, serotonin, and acetylcholine neurons. However, mitochondrial disturbances, inflammatory changes and genetic factors are other potential mechanisms that can contribute to cognitive decline in PD [22]. Biomarkers that will help us with predicting future cognitive decline are needed. There is some evidence showing that low cerebrospinal fluid levels of amyloid-β_{42}, can be predictive for future cognitive decline and dementia in PD [26–28]. Genetic factors like the APOE*ε4 allele, GBA mutations, triplications in the α-synuclein gene are very important for predicting cognitive problems in PD [29–32]. New neuroimaging, especially MRI and PET are also very helpful in early recognition of cognitive decline. Multiple brain regions are suggested and investigated to be involved in cognitive decline PDD patients by neuroimaging methods. MRI studies have shown that cognitive decline in PD is associated with disruption of corticos-triatal and frontal cortex functional connectivity [33, 34] and PET studies with glucose metabolism have shown metabolic decreases in the parietal, temporal, cingulate, and frontal cortices in PD-MCI and PDD [35, 36]. Some reports are showing associations between the slowing of the EEG in PD and cognitive deficits and even a predictive role of EEG slowing for the transition to dementia at 5 years [37, 38]. The treatment options are still not very effective considering PD-MCI and PDD. A treatment that can stop or delay cognitive decline is needed. Using antidementia drugs like cholinesterase inhibitors especially rivastigmine has some effect on PD dementia. Other non-pharmacological methods are cognitive training and physical exercise. There are some investigations in finding good target places for DBS in PDD. One of them is the stimulation of the cholinergic nucleus basalis of Meynert although DBS can decrease slightly executive functions [39, 40].

Depression

Depression is very common in PD patients (about 35%of PD patients) but often untreated [41]. Less than 20% of PD patients with depression are treated [42]. It is associated with poor quality of life, cognitive impairment, functional limitations, caregiver burden, disease duration, use of levodopa in therapy and mortality [43]. Risk factors are also female gender, a history of anxiety or depression, a family history of depression, worse functioning on activities of daily living. Depression together with hyposmia, rapid eye movements sleep behavior disorder (RBD), and constipation can precede the onset of motor symptoms. Depression in PD is also related to changes in dopaminergic, noradrenergic, and serotonergic systems [44]. Some neuroimaging studies have shown frontal lobe atrophy, increased neural activity in the prefrontal regions and decreased functional connectivity between the prefrontal—limbic networks, amygdala, and hippocampus [45, 46]. There are a lot of scales for the detection of depression and should be used routinely. After the recognition, the appropriate management is very important. There are a lot of studies showing that antidepressants, monoamine oxidase-B inhibitors, and dopamine agonists are effective in treating depressive symptoms in PD [47–49]. Repetitive transcranial magnetic stimulation (rTMS) is still in research for depression in PD but is promising [50], like electroconvulsive therapy [51]. A lot of non-pharmacological methods like psychodynamic therapy, exercise, stress release, pet therapy, light therapy, group music, dancing, and singing therapies can help [52]. DBS effect on depression is controversial. Some studies have shown no difference between DBS treating patients and best medical treatment [53], some that there is worsening of depression [54] and some improvement [55].

Anxiety

Anxiety is also very frequent in PD patients (up to 60% of PD patients) and can precede together with depression the onset of motor symptoms [56]. It can be presented as generalized anxiety with fear and worry, panic attacks, and social phobia. Risk factors for anxiety are motor severity, women gender, motor fluctuations, and younger age of onset [57]. A new study has shown that anxiety is associated, also, with higher complications of PD therapy, higher depression, and lower quality of life and it is also poorly recognized and undertreated. Although it contributes to greater disability and worse quality of life, up to 60% of PD patients with anxiety are not treated for anxiety [58]. The pathophysiology and biochemical basis for anxiety in PD are still unknown. There are a lot of scales for anxiety but the

Hamilton Anxiety (HAM-A) rating scale is used the most [59]. Treatment options are benzodiazepines or selective serotonin reuptake inhibitors (SSRIs), cognitive behavioral therapy, and psychotherapy. Subthalamic DBS can improve anxiety in PD patients [60]. In PD patients with anxiety and depression together the treatment should be more aggressive and intensive.

Apathy and Fatigue

Apathy can be present in up to 60% of PD patients with or without depression and dementia [61]. Clinical manifestation of apathy is a loss of motivation, but it also cavers a decrease in goal-directed behavior, cognition and emotions. Nevertheless, there is still no consensus about the definition and validated diagnostic criteria for apathy in PD. Risk factors for apathy in PD are older age, lower cognitive score, higher motor scores and worse activities of daily living function, and poor independency. It is connected with poor quality of life and increased caregiver burden [62]. Some neuroimaging studies have found importance of the frontal, limbic, and striatal connections, marked serotonin denervation in some brain regions in PD patients with apathy [63]. Dopaminergic therapy and DBS can help, but some studies have shown that in the first months after DBS apathy can increase, probably due to a rapid withdrawal of dopaminergic drugs after DBS. However, the influence of greater mesolimbic degeneration, non-motor fluctuations and dyskinesias is very important in increasing the apathy after DBS [64].

Fatigue is also a frequent non-motor symptom in PD patients. Approximately 50% of PD patients complain of fatigue [65]. Patients' complains are lack of energy, exhaustion, and tiredness. It can be present in patients with good motor function or bad motor function. It is often present in patients with sleep problems, depression, and autonomic dysfunction. With decreasing off periods, fatigue can be reduced [66].

The pathophysiology of fatigue in PD is unknown but it is suggested that it is connected with basal ganglia dysfunction. The Fatigue Severity Scale and Parkinson's Fatigue Scale are used for screening of fatigue. Due to the fact that fatigue is one of a major source of disability, regular screening has to be standardized. Known medications for fatigue failed so far in reducing of fatigue in PD patients, but we have to support also non-pharmacological methods like regular exercise and increased physical activity.

Impulse Control Disorders

Impulse control disorder (ICD) includes behavioral disorders with stereotyped, useless and obsessive activities that influence activities of daily life and normal social functioning. It includes hypersexuality, gambling, shopping, punding, and compulsive eating, but the number is growing. Studies have shown that is connected to using dopamine agonists in PD, restless legs syndrome and pro-lactinomas, but some reported it in cases with using levodopa alone. The influence of dopamine D3 activity is suggested as a major role in ICD, although the pathophysiology has to be elucidated [67]. Up to 40% of PD patients treated with dopamine agonists have ICD. Risk factors are younger age, being unmarried, taking levodopa, a family history of ICD or alcoholism and smoking. Shopping is more frequent in women and hypersexuality in men [68]. There are some benefits in ICD reducing antipsychotics, reduction or withdrawal of dopamine agonist, however, the results of DBS studies are controversial. Good results in DBS treated PD patients with ICD are in a hyperdopaminergic condition of patients due to a reduction of the dopaminergic drugs after DBS [69].

Psychosis

Psychosis is a non-motor symptom with usual onset in late phase of drug-treated PD patients, but minor symptoms can be present earlier. It includes mostly visual hallucinations and delusions (up to 40% of PD patients) and rarely auditory, tactile, olfactory, and gustatory hallucinations [70]. Mental state, changing of environment, abnormalities in visual pathways, presences of Lewy bodies in amygdala, frontal, temporal-parietal, and visual cortices, disturbances of serotonin and sleep problems are associated with hallucinations in PD patients. Risk factors are also anticholinergic bladder medications, narcotics, sleep medications, and infectious disorders, therefore, have to be avoided [71]. Delusions are often paranoid and often about feints of family members, especially spouse. Antiparkinsonian drugs should be reduced, first anticholinergics, then amantadin, and dopamine agonist. If this is not enough, we can reduce monoamine oxidase-B inhibitors and in the end levodopa. Pimavanserin is a new drug for PD psychosis. Other possibilities are clozapine and quetiapine.

Future Perspective

Recent studies have shown that pump-based Parkinson (PD) therapies, including subcutaneous apomorphine infusion (CSA) and levodopa-carbidopa intestinal gel (LCIG) may improve sleep, mood, and apathy, gastrointestinal symptoms, and urological symptoms. However, they can worsen some NMS [72, 73]. Knowing effects on NMS of all available medications and invasive treatments (DBS, CSA, and LCIG) is very important in decision-making which invasive treatment to choose [74]. This is the first step to the personalized management of PD, which is with precision medicine and pharmacogenetics the future in PD treatment [75]. The update of NMS treatment options is recently published and has to be regularly published to provide clinicians and investigators with an up-to-date evidence base for better management of NMS in PD [76].

Conclusions

Earlier recognition of non-motor symptoms especially neuropsychiatric features is needed for prompt intensive treatment that will lead to improvement of quality of life PD patients and their caregivers. Early recognition and effective treatment of NMS will be some of the most challenging achievements for clinicians and researchers in the PD field. The personalized treatment of PD involves holistic, modern treatment of motor and non-motor symptoms and future studies have to bring us more reliable evidence of the best treatment options for that.

References

1. Schapira AH, Tolosa E. Molecular and clinical prodrome of Parkinson disease: implications for treatment. Nat Rev Neurol. 2010;6:309–17. https://doi.org/10.1038/nrneurol.2010.52.
2. Klingelhoefer L, Samuel M, Chaudhuri KR, Ashkan K. An update of the impact of deep brain stimulation on non- motor symptoms in Parkinson's disease. J Parkinsons Dis. 2014;4:289–300. https://doi.org/10.3233/JPD-130273.
3. Seppi K, et al. The movement disorder society evidence-based medicine review update: treatments for the non-motor symptoms of Parkinson's disease. Mov Disord. 2011;26(Suppl. 3):S42–80. https://doi.org/10.1002/mds.23884.
4. Hindle JV. Ageing, neurodegeneration and Parkinson's disease. Age Ageining. 2010;39(2):156–611. https://doi.org/10.1093/ageing/afp223.
5. Chaudhuri KR, Healy DG, Schapira AH. Non-motor symptoms of Parkinson's disease: diagnosis and management. Lancet Neurol. 2006;5:235–45. https://doi.org/10.1016/S1474-4422(06)70373-8.
6. Aaarslan D, Larsen JP, Tandberg E, Laake K. Predictors of nursing home placement in Parkinson's disease: a population-based prospective study. J Am Geriatr Soc. 2000;48:938–42. https://doi.org/10.1111/j.1532-5415.2000.tb06891.
7. Chaudhuri KR, et al. International multicenter pilot study of the first comprehensive self-completed nonmotor symptoms questionnaire for Parkinson's disease: the NMSQuest study. Mov Disord. 2006;21:916–23. https://doi.org/10.1002/mds.20844.
8. Chaudhuri KR, et al. The metric properties of a novel non-motor symptoms scale for Parkinson's disease: results from an international pilot study. Mov Disord. 2007;22:1901–11. https://doi.org/10.1002/mds.21596.
9. Chaudhuri KR, et al. King's Parkinson's disease pain scale, the first scale for pain in PD: an international validation. Mov Disord. 2015;30:1623–31. https://doi.org/10.1002/mdc3.12384.
10. Chaudhuri KR, Martinez-Martin P. Clinical assessment of nocturnal disability in Parkinson's disease: the Parkinson's disease Sleep Scale. Neurology. 2004;63:S17–20. https://doi.org/10.1212/wnl.63.8_suppl_3.s17.
11. Grinberg LT, Rueb U, Alho AT, Heinsen H. Brainstem pathology and non-motor symptoms in PD. J Neurol Sci. 2010;289:81–8. https://doi.org/10.1016/j.jns.2009.08.021.
12. Hely MA, Reid WG, Adena MA, Halliday GM, Morris JG. The Sydney multicenter study of Parkinson's disease: the inevitability of dementia at 20 years. Mov Disord. 2008;23:837–44. https://doi.org/10.1002/mds.21956.
13. EmreM, Aarsland D, Brown R, et al. Clinical diagnostic criteria for dementia associated with Parkinson's disease. Mov Disord. 2007;22(12):1689–1707. https://doi.org/10.1002/mds.21507.

14. Pigott K, Rick J, Xie SX, et al. Longitudinal study of normal cognition in Parkinson disease. Neurology. 2015;85(15):1276–82. https://doi.org/10.1212/WNL.0000000000002001.
15. Cereda E, Cilia R, Klersy C, et al. Dementia in Parkinson's disease: is male gender a risk factor? Parkinsonism Relat Disord. 2016;26:67–72. https://doi.org/10.1016/j.parkreldis.2016.02.024.
16. Anang JB, Gagnon J-F, Bertrand J-A, et al. Predictors of dementia in Parkinson disease: a prospective cohort study. Neurology. 2014;83(14):1253–60. https://doi.org/10.1212/WNL.0000000000000842.
17. Kempster PA, O'Sullivan SS, Holton JL, Revesz T, Lees AJ. Relationships between age and late progression of Parkinson's disease: a clinico-pathological study. Brain. 2010;133(Pt 6):1755–62. https://doi.org/10.1093/brain/awq059.
18. Monchi O, Hanganu A, Bellec P. Markers of cognitive decline in PD: The case for heterogeneity. Parkinsonism Relat Disord. 2016;24:8–14. https://doi.org/10.1016/j.parkreldis.2016.01.002.
19. Williams-Gray CH, Evans JR, Goris A, et al. The distinct cognitive syndromes of Parkinson's disease: 5 year follow-up of the Cam-PaIGN cohort. Brain. 2009;132(Pt 11):2958–69. https://doi.org/10.1093/brain/awp245.
20. Svenningsson P, Westman E, Ballard C, Aarsland D. Cognitive impairment in patients with Parkinson's disease: diagnosis, biomarkers, and treatment. Lancet Neurol. 2012;11:697–707. https://doi.org/10.1038/nrneurol.2017.27.
21. Litvan I, Goldman JG, Troster AI, et al. Diagnostic criteria for mild cognitive impairment in Parkinson's disease: movement disorder society task force guidelines. Mov Disord. 2012;27:349–56. https://doi.org/10.1002/mds.24893.
22. Halliday GM, Leverenz JB, Schneider JS, Adler CH. The neurobiological basis of cognitive impairment in Parkinson's disease. Mov Disord. 2014;29:634–50. https://doi.org/10.1002/mds.25857.
23. Braak H, et al. Staging of brain pathology related to sporadic Parkinson's disease. Neurobiol Aging. 2003;24:197–211. https://doi.org/10.1016/s0197-4580(02)00065-9.
24. Compta Y, et al. Lewy- and Alzheimer-type pathologies in Parkinson's disease dementia: which is more important? Brain. 2011;134:1493–505. https://doi.org/10.1093/brain/awr031.
25. Howlett DR, et al. Regional multiple pathology scores are associated with cognitive decline in Lewy body dementias. Brain Pathol. 2015;25:401–8. https://doi.org/10.1111/bpa.12182.
26. Backstrom DC, et al. Cerebrospinal fluid patterns and the risk of future dementia in early, incident Parkinson disease. JAMA Neurol. 2015;72:1175–82. https://doi.org/10.1001/jamaneurol.2015.1449.
27. Hall S, et al. CSF biomarkers and clinical progression of Parkinson disease. Neurology. 2015;84:57–63. https://doi.org/10.1001/jamaneurol.2015.1449.
28. Alves G, et al. CSF Aβ42 predicts early-onset dementia in Parkinson disease. Neurology. 2014;82:1784–90. https://doi.org/10.1212/WNL.0000000000000425.
29. Guella I, et al. α-Synuclein genetic variability: a biomarker for dementia in Parkinson disease. Ann Neurol. 2016;79:991–9. https://doi.org/10.1002/ana.24664.
30. Alcalay RN, et al. Cognitive performance of GBA mutation carriers with early-onset PD: the CORE-PD study. Neurology. 2012;78:1434–40. https://doi.org/10.1212/WNL.0b013e318253d54b.
31. Williams-Gray CH, et al. Apolipoprotein E genotype as a risk factor for susceptibility to and dementia in Parkinson's disease. J Neurol. 2009;256:493–8. https://doi.org/10.1007/s00415-009-0119-8.
32. Morley JF, et al. Genetic influences on cognitive decline in Parkinson's disease. Mov Disord. 2012;27:512–8. https://doi.org/10.1007/s00415-009-0119-8.
33. Seibert TM, Murphy EA, Kaestner EJ, Brewer JB. Interregional correlations in Parkinson disease and Parkinson-related dementia with resting functional MR imaging. Radiology. 2012;263:226–34. https://doi.org/10.1148/radiol.12111280.

34. Rektorova I, Krajcovicova L, Marecek R, Mikl M. Default mode network and extrastriate visual resting state network in patients with Parkinson's disease dementia. Neurodegener Dis. 2012;10:232–7. https://doi.org/10.1002/gps.4342.
35. Vander Borght T, et al. Cerebral metabolic differences in Parkinson's and Alzheimer's diseases matched for dementia severity. J Nucl Med. 1997;38:797–802.
36. Gonzalez-Redondo R, et al. Grey matter hypometabolism and atrophy in Parkinson's disease with cognitive impairment: a two-step process. Brain. 2014;137:2356–67. https://doi.org/10.1093/brain/awu159.
37. Kamei S, Morita A, Serizawa K, Mizutani T, Hirayanagi K. Quantitative EEG analysis of executive dysfunction in Parkinson disease. J Clin Neurophysiol. 2010;27:193–7. https://doi.org/10.1097/WNP.0b013e3181dd4fdb.
38. Klassen BT, et al. Quantitative EEG as a predictive biomarker for Parkinson disease dementia. Neurology. 2011;77:118–24. https://doi.org/10.1212/WNL.0b013e318224af8d.
39. Gratwicke J, et al. The nucleus basalis of Meynert: a new target for deep brain stimulation in dementia? Neurosci Biobehav Rev. 2013;37:2676–88. https://doi.org/10.1016/j.neubiorev.2013.09.003.
40. Parsons TD, Rogers SA, Braaten AJ, et al. Cognitive sequelae of subtalamic nucleus deep brain stimulation in Parkinson's disease: a meta-analysis. Lancet Neurol. 2006;5:578–88. https://doi.org/10.1016/S1474-4422(06)70475-6.
41. Reijnders JS, Ehrt U, Weber WE, Aarsland D, Leentjens AF. A systematic review of prevalence studies of depression in Parkinson's disease. Mov Disord. 2008;23:183–9. https://doi.org/10.1002/mds.21803.
42. Frisina PG, Borod JC, Foldi NS, Tenenbaum HR. Depression in Parkinson's disease: health risks, etiology, and treatment options. Neuropsychiatr Dis Treat. 2008;4:81–91.
43. Rocha FL, Murad MG, Stumpf BP, Hara C, Fuzikawa C. Antidepressants for depression in Parkinson's disease: systematic review and meta-analysis. J Psychopharmacol. 2013;27:417–23. https://doi.org/10.1177/0269881113478282.
44. Remy P, Doder M, Lees A, Turjanski N, Brooks D. Depression in Parkinson's disease: loss of dopamine and noradrenaline innervation in the limbic system. Brain. 2005;128:1314–22. https://doi.org/10.1093/brain/awh445.
45. Henderson R, Kurlan R, Kersun JM, Como P. Preliminary examination of the comorbidity of anxiety and depression in Parkinson's disease. J Neuropsychiatry Clin Neurosci. 1992;4(03):257–64. https://doi.org/10.1176/jnp.4.3.257.
46. Leentjens AF, Dujardin K, Marsh L, Martinez-Martin P, Richard IH, Starkstein SE. Anxiety and motor fluctuations in Parkinson's disease: a cross-sectional observational study. Parkinsonism Relat Disord. 2012;18(10):1084–8. https://doi.org/10.1016/j.parkreldis.2012.06.007.
47. Troeung L, Egan SJ, Gasson N. A meta-analysis of randomised placebo-controlled treatment trials for depression and anxiety in Parkinson's disease. PLoS ONE. 2013;8:e79510. https://doi.org/10.1371/journal.pone.0079510.
48. Leentjens AF, Koester J, Fruh B, Shephard DT, Barone P, Houben JJ. The effect of pramipexole on mood and motivational symptoms in Parkinson's disease: a meta-analysis of placebo-controlled studies. Clin Ther. 2009;31:89–98. https://doi.org/10.1016/j.clinthera.2009.01.012.
49. Barone P, Santangelo G, Morgante L, et al. A randomized clinical trial to evaluate the effects of rasagiline on depressive symptoms in non-demented Parkinson's disease patients. Eur J Neurol. 2015;22:1184–91. https://doi.org/10.1111/ene.12724.
50. Xie C-L, Chen J, Wang X-D, et al. Repetitive transcranial magnetic stimulation (rTMS) for the treatment of depression in Parkinson disease: a meta-analysis of randomized controlled clinical trials. Neurol Sci. 2015;36(10):1751–61. https://doi.org/10.1007/s10072-015-2345-4.

51. Borisovskaya A, Bryson WC, Buchholz J, Samii A, Borson S. Electroconvulsive therapy for depression in Parkinson's disease: systematic review of evidence and recommendations. Neurodegener Dis Manag. 2016;6(02):161–76. https://doi.org/10.2217/nmt-2016-0002.

52. Xie CL, Wang XD, Chen J, et al. A systematic review and meta-analysis of cognitive behavioral and psychodynamic therapy for depression in Parkinson's disease patients. Neurol Sci. 2015;36:833–43. https://doi.org/10.1007/s10072-015-2118-0.

53. Deuschl G, Schade-Brittimger C, Krak P, et al. A randomized trial of deep-brain stimulation for Parkinson's disease. N Engl J Med. 2006;355:896–908. https://doi.org/10.1056/NEJMoa060281.

54. Follett KA, et al. Pallidal versus subthalamic deep-brain stimulation for Parkinson's disease. N Engl J Med. 2010;362:2077–91. https://doi.org/10.2147/NDT.S105513.

55. Casteli L, et al. Chronic deep brain stimulation of the subthalamic nucleus for Parkinson's disease: effect on cognition, mood, anxiety an personality traits. Eur Neurol. 2006;55:136–44. https://doi.org/10.1159/000093213.

56. Lin CH, Lin JW, Liu YC, Chang CH, Wu RM. Risk of Parkinson's disease following anxiety disorders: a nationwide population-based cohort study. Eur J Neurol. 2015;22:1280–7. https://doi.org/10.1111/ene.12740.

57. Brown RG, et al. Depression and anxiety related subtypes in Parkinson's disease. J Neurol Neurosurg Psychiatry. 2011;82:803–9.

58. Dissanayaka NNW, White E, O'Sullivan JD, Marsh R, Silburn PA, Copland DA, Mellick GD, Byrne GJ. Characteristics and Treatment of Anxiety Disorders in Parkinson's Disease. Mov Disord Clin Pract. 2015;2:155–62.

59. https://doi.org/10.1136/jnnp.2010.213652.

60. Dissanayaka NNW, O'Sullivan JD, Silburn PA, Mellick GD. Assessment methods and factors associated with depression in Parkinson's disease. J Neurol Sci. 2011;310:208–10. https://doi.org/10.1016/j.jad.2011.01.021.

61. Witt K, et al. Neuropsychological and psychiatric changes after deep brain stimulation for Parkinson's disease: a randomised, multicentre study. Lancet Neurol. 2008;7:605–14. https://doi.org/10.1016/S1474-4422(08)70114-5.

62. Pedersen KF, et al. Apathy in drug-naive patients with incident Parkinson's disease: the Norwegian Park West study. J Neurol. 2010;257:217–23. https://doi.org/10.1007/s00415-009-5297-x.

63. den Brok MG, van Dalen JW, van Gool WA, Moll van Charante EP, de Bie RM, Richard E. Apathy in Parkinson's disease: a systematic review and meta-analysis. Mov Disord 2015;30(06): 759–69. https://doi.org/10.1002/mds.26208.

64. Maillet A, Krack P, Lhommée E, et al. The prominent role of serotonergic degeneration in apathy, anxiety and depression in de novo Parkinson's disease. Brain. 2016;139(Pt 9):2486–502. https://doi.org/10.1093/brain/aww162.

65. Thobois S, et al. Non-motor dopamine withdrawal syndrome after surgery for Parkinson's disease: predictors and underlying mesolimbic denervation. Brain. 2010;133:1111–27. https://doi.org/10.1093/brain/awq032.

66. Del SF, Albanese A. Clinical management of pain and fatigue in Parkinson's disease. Parkinsonism Relat Disord. 2012;18(Suppl. 1):S233–6. https://doi.org/10.1016/S1353-8020(11)70071-2.

67. Hagell P, Brundin L. Towards an understanding of fatigue in Parkinson disease. J Neurol Neurosurg Psychiatry. 2009;80:489–92. https://doi.org/10.1136/jnnp.2008.159772.

68. Moore TJ, Glenmullen J, Mattison DR. Reports of pathological gambling, hypersexuality, and compulsive shopping associated with dopamine receptor agonist drugs. JAMA Intern Med. 2014;174(12):1930–3. https://doi.org/10.1001/jamainternmed.2014.5262.

69. Evans AH, Lawrence AD, Potts J, Appel S, Lees AJ. Factors influencing susceptibility to compulsive dopaminergic drug use in Parkinson disease. Neurology. 2005;65(10):1570–4. https://doi.org/10.1212/01.wnl.0000184487.72289.f0.

70. Shotbolt P, et al. Relationships between deep brain stimulation and impulse control disorders in Parkinson's disease, with a literature review. Parkinsonism Relat. Disord. 2012;18:10–6. https://doi.org/10.1016/j.parkreldis.2011.08.016.
71. Zahodne LB, Fernandez HH. Pathophysiology and treatment of psychosis in Parkinson's disease: a review. Drugs Aging. 2008;25:665–82. https://doi.org/10.2165/00002512-200825080-00004.
72. Papapetropoulos S, McCorquodale DS, Gonzalez J, Jean-Gilles L, Mash DC. Cortical and amygdalar Lewy body burden in Parkinson's disease patients with visual hallucinations. Parkinsonism Relat Disord. 2006;12:253–6. https://doi.org/10.1016/j.parkreldis.2005.10.005.
73. Mundt-Petersen U, Odin P. Infusional therapies, continuous dopaminergic stimulation, and nonmotor symptoms. In: Chaudhuri R, Titova N, editors, Parkinson's: the hidden face management and the hidden face of related disorders. Elsevier. 2017. p. 1019–44. (International Review of Neurobiology). https://doi.org/10.1016/bs.irn.2017.05.036.
74. Ray Chaudhuri K, Antonini A, Robieson WZ, Sanchez-Solino O, Bergmann L, Poewe W, et al. Burden of non-motor symptoms in Parkinson's disease patients predicts improvement in quality of life during treatment with levodopa-carbidopa intestinal gel. Eur J Neurol. 2019;26(4):581–e43. https://doi.org/10.1111/ene.13847.
75. Dafsari HS, Martinez-Martin P, Rizos A, Trost M, Dos Santos Ghilardi MG, Reddy P, et al. EuroInf 2: subthalamic stimulation, apomorphine, and levodopa infusion in Parkinson's disease. Mov Disord. 2019;34:353–65. https://doi.org/10.1002/mds.27626.
76. Titova N, Chaudhuri KR. Non-motor Parkinson disease: new concepts and personalised management. Med J Australia. 2018;208(9):404–9. https://doi.org/10.5694/mja17.00993.
77. Seppi K, Ray Chaudhuri K, Coelho M, Fox SH, Katzenschlager R, Perez Lloret S, Weintraub D, Sampaio C. The collaborators of the Parkinson's disease update on non-motor symptoms study group on behalf of the movement disorders society evidence-based medicine committee update on treatments for nonmotor symptoms of Parkinson's disease-an evidence-based medicine review. Mov Disord. 2019;34:180–98. https://doi.org/10.1002/mds.27602.

Vascular Cognitive Impairment

Petra Črnac Žuna, Hrvoje Budinčević, Tena Sučić Radovanović, Milija Mijajlović and Natan Bornstein

Introduction

Dementia is a syndrome characterized by the deterioration of cognitive functions, beyond what might be expected from normal aging, leading to the disruption of the ability to perform daily activities [1]. Dementia is one of the main causes of disability and personal care assistance requirements in elderly people throughout the world [2]. The prevalence of dementia is 5–10% in people over 65, and the prevalence of vascular dementia doubles every 5.3 years [3].

Clinically we can distinguish cortical and subcortical dementia. Cortical dementia is characterized by aphasia, agnosia, apraxia, and amnesia with impaired executive function (planning, organization, judgment). The prototype for this type of dementia is Alzheimer's dementia (AD). For instance, in different cortical types of dementia, patients with medial frontal lobe lesions present with executive function impairment, abulia, and/or apathy. Patients with bilateral frontal damage present with akinetic mutism. Patients with left parietal lobe lesions present with aphasia, apraxia and agnosia, while right parietal lobe lesions cause neglect (anosognosia, asomatognosia), confusion, agitation.

P. Črnac Žuna · H. Budinčević (✉)
Department of Neurology, Sveti Duh University Hospital, Zagreb, Croatia
e-mail: hbudincevic@gmail.com

H. Budinčević
Faculty of Medicine, Josip Juraj Strossmayer University of Osijek, Osijek, Croatia

T. Sučić Radovanović
Department of Radiology, Sveti Duh University Hospital, Zagreb, Croatia

M. Mijajlović
Clinical Center of Serbia and School of Medicine University of Belgrade, Neurology Clinic, Belgrade, Serbia

N. Bornstein
Brain Division, Shaare Zedek Medical Center, Jerusalem, Israel

© Springer Nature Switzerland AG 2020
V. Demarin (ed.), *Mind and Brain*, https://doi.org/10.1007/978-3-030-38606-1_10

In patients with medial temporal lobe lesions, anterograde amnesia can be expected [4]. Subcortical dementia is characterized primarily by bradyphrenia (decline in mental processing speed) and impairment of procedural memory, planning and reasoning, with associated focal motor signs, gait and urination disorders, pseudobulbar paralysis, and emotional incontinence [5]. The prototype of subcortical dementia is vascular dementia. In vascular dementia, executive functions are usually impaired and milder memory impairment maybe present [6]. The onset of vascular dementia can be sudden, usually after strokes, but more often the course is progressive with worsening of symptoms after each new vascular event [7].

Traditionally, the most common type of dementia is considered to be Alzheimer's dementia, the cause of all chronic dementia in 50–75% of cases [8]. Mild cognitive impairment is a transitional form between normal cognitive status and Alzheimer's dementia [9]. Vascular dementia is the second most common type of dementia [10]. It is a result of cerebrovascular disease and is thought to be the cause of 20–30% of all chronic dementias [11]. Men are more likely to be affected by vascular dementia than women [12].

10% of patients already have dementia at stroke onset. 10% of patients will develop dementia after a first-ever stroke. One-third of patients will develop dementia with stroke recurrence [13].

In total, post-stroke dementia (PSD) or post-stroke cognitive impairment (PSCI) may affect up to one-third of stroke survivors [14]. The transitional form between normal cognitive status and vascular dementia is called vascular cognitive impairment (VCI) [15]. VCI may be the most preventable and treatable cause of dementia [16].

In general, 90% of strokes and 35% of dementias have been estimated to be preventable. A stroke doubles the chance of developing dementia and stroke is more common than dementia, therefore, more than one-third of dementia could be prevented by preventing stroke [17].

More recent imaging and clinical-pathological studies demonstrated that ischemic lesions, neurovascular dysfunction, and AD pathology most often coexist in the same brain, and verified destructive cerebrovascular effects of amyloid-beta, so "mixed dementia" maybe the most common cause of cognitive impairment in the elderly, after all [18].

Cognitive symptoms in vascular cognitive impairment are characterized by psychomotor slowing, complex attention deficits, executive function and memory retrieval deficits. On the contrary, cognitive symptoms in Alzheimer's disease are related to short-term memory deficits, word-finding difficulties, visuospatial and memory encoding deficits [17]. Vascular dementia is usually associated with apathy, depression or hallucinations, and delirium. On the other hand, in Alzheimer's disease delusions and loss of insight are more common [17]. Contrary to Alzheimer's disease, vascular dementia is often associated with focal neurological signs or parkinsonism [17].

Pathophysiology

In the pathophysiology of vascular cognitive impairment, the function of the neurovascular unit is impaired, including neurons, glial cells, perivascular, and vascular structures [19]. The accumulation of beta-amyloid has an important role in this process: it is a powerful vasoconstrictor, participating in the development of cerebral amyloid angiopathy [20]. In addition, lipohyalinosis, reactive astrocytosis, and microglial activation play an important role in pathophysiology [21]. Oxidative stress and inflammation induced by cardiovascular risk factors and Aβ (amyloid-beta) are responsible for the impairment of the neurovascular unit functions, leading to local hypoxia–ischemia, axonal demyelination, and decreased repair potential of the white matter [22].

Myelin loss increases energy consumption and increases local hypoxia [23].

In vascular dementia, cerebrovascular risk factors induce neurovascular dysfunction, leading to cerebrovascular insufficiency, resulting in brain dysfunction. Aβ induces vascular dysregulation and aggravates the vascular insufficiency, worsening the brain dysfunction associated with vascular risk factors [24].

On the other hand, the hypoxia–ischemia caused by vascular insufficiency increases Aβ cleavage from amyloid precursor protein and diminishes Aβ clearance, promoting Aβ accumulation and the resulting nocuous effects [25].

Given the vascular genesis of the disease, most of the risk factors overlap with risk factors for stroke and coronary artery disease, such as age, physical inactivity, obesity, smoking, inappropriate diet, excessive alcohol consumption, arterial hypertension, diabetes, peripheral arterial disease, chronic kidney disease and coronary artery disease, low cardiac output, and atrial fibrillation [26]. Additional overlapping risk factors with Alzheimer's disease are low levels of education and social isolation [27].

Subclassification of Vascular Cognitive Impairment

VCI can be further classified into two forms: post-stroke and non-stroke related [28]. VCI can occur as a result of a single stroke in strategic locations, particularly if the following regions are affected: the angular gyrus, left hemisphere perisylvian language areas, the mediodorsal part of the temporal lobe and the anterior thalamus, the midbrain, the medial frontal lobe [29]. VCI may also result from the joined effect of multiple strokes involving sufficient brain regions, for instance, the prefrontal lobes, participating in cognitive processing (previously known as multi-infarct dementia) [30]. However, VCI may also be non-stroke related, as a result of diffuse, severe subcortical cerebrovascular disease as part of microangiopathy, cerebral small vessel disease, also known as Binswanger's disease [31].

Classification Criteria

The classification criteria for VCI have been published by the AHA/ASA, Vas-Cog, DSM-5, ICD-10, NINDS-AIREN, and ADDTC: all of them require some extent of cognitive impairment should be present, as well as affirmation of a vascular contribution to the cognitive impairment [31–36].

As reported by the more recent classification system, we can differentiate the full spectrum of cognition; from vascular mild cognitive impairment (or minor vascular neurocognitive disorder) and vascular dementia (or major vascular neurocognitive disorder) [31, 33].

VCI can be further classified as probable or possible, based on the presence or absence of indicators for competing causes of dementia (i.e. Alzheimer disease), or the completeness of the diagnostic examination [31, 33].

Neuroimaging has determined that clinically silent, "covert" cerebrovascular disease becomes common with aging and is sometimes sufficient to cause cognitive impairment, therefore, the non-existence of a clinical history of stroke does not exclude VCI [37].

Vascular pathologies underlying vascular cognitive impairment are extremely diverse, therefore, management and prognosis must be highly individualized [31].

All patients with cognitive impairment should be evaluated for contributing to vascular causes. The diagnostic process consists of three stages: determination of cognitive impairment; presence, severity, and cause of vascular disease; indicators of vascular contribution to the cognitive impairment [28].

Cognitive Evaluation

To assess the presence and the severity of cognitive impairment, we can use the following rating scales: (1) the Mini Mental Status Examination (MMSE) [38], (2) the Montreal Cognitive Assessment (MoCA) [39]—including the clock drawing test (Clock Drawing Test (CDT) [40] and (3) the Hachinski Ischemic Score (HIS) [41]. Useful bedside screening tools sensitive for VCI detection should assess primarily processing speed and executive function, as they manifest with relatively greater impairments than episodic memory: MoCA appears to be more sensitive than the MMSE in the assessment of VCI [42].

Assessment of Cerebrovascular Disease

A synthesis of history, examination, and neuroimaging are fundamental for the assessment of the presence of cerebrovascular disease. The clearest evidence of cerebrovascular disease is when a clinical history of stroke exists, or the physical examination shows focal neurologic signs. On the other hand, neuroimaging identifies evidence of silent strokes; clinical signs and symptoms may also be suggestive of non-stroke VCI (stepwise progressions of cognitive impairment, frontal gait disorder, urinary incontinence, executive dysfunction, and slowed processing speed, minor asymmetric neurologic signs such as increased tone, Babinski responses, reflex asymmetry, frontal release signs), the presence of vascular risk factors should increase suspicion for VCI [31]. Although, each of the abovementioned signs individually should not be taken as proof of VCI, only as a possible contributing factor to strengthen the clinical judgment that the vascular disease is causing the cognitive impairment.

The role of neuroimaging, preferably MRI, is particularly important in the diagnosis of VCI, to confirm the presence of cerebrovascular disease, to assess the severity of the disease and the location of the cerebrovascular lesions [43]. This is important to determine the clinical significance of the lesions and whether they are sufficient to account for cognitive impairment. The diagnosis of probable vascular cognitive impairment requires confirmation of cognitive impairment, cerebrovascular disease, and a clear relationship between the two conditions. AHA/ASA, NINDS-AIREN, and DSM-5 diagnostic criteria are based on this principle [32–34, 42]. Nevertheless, other causes of dementia should be ruled out.

Assessment of the Relationship Between Cerebrovascular Disease and Cognitive Impairment

Assessing the relationship between cerebrovascular disease and cognitive impairment is probably the most clinically challenging part. Clinical judgment is needed to determine whether cerebrovascular lesions are sufficient to cause cognitive impairment on a case-by-case basis because the cerebrovascular disease is relatively common and can be incidental. It is important not to overlook a history of pre-stroke cognitive decline, which could indicate a competing cause of cognitive impairment.

Exact thresholds of lesion number and volume are not well determined and likely vary depending on the location [33]. Also, the clinician must keep in mind that neuroimaging does not have perfect sensitivity; it is unable to detect microinfarcts (infarcts as small as 0.2 mm in diameter) [44].

In addition, the clinical signs of VCI can manifest with other motor and noncognitive manifestations of cerebrovascular disease (apathy, depression, urinary incontinence, frontal gait disorder, lower-body parkinsonism, spasticity, hyperreflexia, frontal release signs) and behavioral disturbances (emotional incontinence, apathy, depression) [31].

Other Causes of Vascular Cognitive Impairment

It is also important to keep in mind nonatherosclerotic and nonarteriolosclerotic causes of VCI: Cerebral amyloid angiopathy and monogenetic inherited causes of VCI.

Cerebral amyloid angiopathy is caused by vascular beta-amyloid deposition in the cerebral cortex and leptomeninges, causing fragility of the vessel wall with rupture and bleeding in some patients. It is the second most common cause of intracerebral bleeding in elderly patients, located primarily in the superficial, lobar portions of the cortex, and underlying white matter, or in the subarachnoid space [45].

Monogenetic inherited causes of VCI should be suspected when the existence of confluent white matter hyperintensities or multiple lacunae are remarkably greater than that expected by age and can't be explained by traditional vascular risk factors [46]. The best-studied inherited cause of VCI is CADASIL (cerebral autosomal dominant arteriopathy with subcortical infarcts and leukoencephalopathy).

Management

Management of VCI includes patient and caregiver support, cognitive rehabilitation of post-stroke VCI including cognitive skills training or training for compensatory strategies, cognitive-enhancing medications, treatment, and secondary prevention and probably the most significant part of management, to prevent the recurrence and progression of the causative cerebrovascular processes [33]. Recent studies show that drugs used to treat Alzheimer disease, such as donepezil and memantine show the modest benefit in patients with VCI, if any, on standard cognitive measures: small samples of executive dysfunctions, inconsistent benefit in global and daily function, not distinguishable from Alzheimer's disease [33].

Treatment of VCI should be aimed at treating vascular risk factors (arterial hypertension, diabetes, hyperlipidemia) and prevention of stroke [47].

Hypertension is the strongest risk factor for stroke overall [48]. Patients with cerebral small vessel disease should have a measurement of blood pressure, ECG to assess for atrial fibrillation, serum lipid profile, and blood glucose measurement. The more extensive evaluation includes assessment for proximal sources of embolism (echocardiography, prolonged cardiac rhythm monitoring, noninvasive carotid imaging).

There are no proven preventive measures for cognitive impairment in patients with cerebral small vessel disease [49]. In this subset of patients is reasonable to start aspirin, and the use of statins may be considered on a case-by-case basis. Good blood pressure control probably slows the progression of white matter hyperintensity [50].

According to the FINGER study, effective prevention includes improved population control of vascular risk factors: multidomain intervention, including diet, exercise, and cognitive training prevented a decline in cognitive test scores [51]. Lifestyle modification with adequate nutrition and hydration, physical activity, smoking cessation, weight control, and moderate alcohol consumption should certainly be an integral part of the prevention and treatment of vascular cognitive impairment [28].

References

1. Bowler JV. Vascular cognitive impairment. J Neurol Neurosurg Psychiatry. 2005;76(Suppl 5): v35–44. Epub 2005/11/18.
2. Yoshida D, Ninomiya T, Doi Y, Hata J, Fukuhara M, Ikeda F, et al. Prevalence and causes of functional disability in an elderly general population of Japanese: the Hisayama study. J Epidemiol. 2012;22(3):222–9. Epub 2012/02/22.
3. Jorm AF, Korten AE, Henderson AS. The prevalence of dementia: a quantitative integration of the literature. Acta Psychiatr Scand. 1987;76(5):465–79. Epub 1987/11/01.
4. Patterson C, Gauthier S, Bergman H, Cohen C, Feightner JW, Feldman H, et al. The recognition, assessment and management of dementing disorders: conclusions from the Canadian Consensus Conference on Dementia. Can J Neurol Sci. 2001;28 Suppl 1:S3–16. Epub 2001/03/10.
5. Bowler JV. Modern concept of vascular cognitive impairment. Br Med Bull. 2007;83:291–305. Epub 2007/08/07.
6. Salihovic D, Smajlovic D, Mijajlovic M, Zoletic E, Ibrahimagic OC. Cognitive syndromes after the first stroke. Neurol Sci: Off J Ital Neurol Soc Ital Soc Clin Neurophysiol. 2018;39(8):1445–51. Epub 2018/05/21.
7. Lee AY. Vascular dementia. Chonnam Med J. 2011;47(2):66–71. Epub 2011/11/24.
8. Cunningham EL, McGuinness B, Herron B, Passmore AP. Dementia. Ulst Med J. 2015; 84(2):79–87. Epub 2015/07/15.
9. Petersen RC. Early diagnosis of Alzheimer's disease: is MCI too late? Curr Alzheimer Res. 2009;6(4):324–30. Epub 2009/08/20.
10. Kalaria RN. Neuropathological diagnosis of vascular cognitive impairment and vascular dementia with implications for Alzheimer's disease. Acta Neuropathol. 2016;131(5):659–85. Epub 2016/04/12.
11. Iadecola C. The pathobiology of vascular dementia. Neuron. 2013;80(4):844–66. Epub 2013/11/26.
12. Ruitenberg A, Ott A, van Swieten JC, Hofman A, Breteler MM. Incidence of dementia: does gender make a difference? Neurobiol Aging. 2001;22(4):575–80. Epub 2001/07/11.
13. Casolla B, Leys D. Stroke and dementia. In: Brainin M, Heiss W-D, editors. Textbook of stroke medicine, 2nd ed. Cambridge University Press; 2014. p. 255–65.
14. Mijajlovic MD, Pavlovic A, Brainin M, Heiss WD, Quinn TJ, Ihle-Hansen HB, et al. Poststroke dementia—a comprehensive review. BMC Med. 2017;15(1):11. Epub 2017/01/18.
15. Demarin V, Basic Kes V, Trkanjec Z, Budisic M, Bosnjak Pasic M, Crnac P, et al. Efficacy and safety of Ginkgo biloba standardized extract in the treatment of vascular cognitive impairment: a randomized, double-blind, placebo-controlled clinical trial. Neuropsychiatr Dis Treat. 2017;13:483–90. Epub 2017/03/01.
16. Roman GC. Vascular dementia may be the most common form of dementia in the elderly. J Neurol Sci. 2002;203–204:7–10. Epub 2002/11/06.
17. Hachinski V, Einhaupl K, Ganten D, Alladi S, Brayne C, Stephan BCM, et al. Preventing dementia by preventing stroke: The Berlin Manifesto. Alzheimer's Dement J Alzheimer's Assoc. 2019;15(7):961–84. Epub 2019/07/23.

18. Schneider JA, Arvanitakis Z, Bang W, Bennett DA. Mixed brain pathologies account for most dementia cases in community-dwelling older persons. Neurology. 2007;69(24):2197–204. Epub 2007/06/15.
19. Kalaria RN. The pathology and pathophysiology of vascular dementia. Neuropharmacology. 2018;134(Pt B):226–39. Epub 2017/12/24.
20. Niwa K, Porter VA, Kazama K, Cornfield D, Carlson GA, Iadecola C. A beta-peptides enhance vasoconstriction in cerebral circulation. Am J Physiol Heart Circ Physiol. 2001;281(6):H2417–24. Epub 2001/11/16.
21. Olsson B, Hertze J, Lautner R, Zetterberg H, Nagga K, Hoglund K, et al. Microglial markers are elevated in the prodromal phase of Alzheimer's disease and vascular dementia. J Alzheimer's Dis: JAD. 2013;33(1):45–53. Epub 2012/08/15.
22. Iadecola C. The overlap between neurodegenerative and vascular factors in the pathogenesis of dementia. Acta Neuropathol. 2010;120(3):287–96. Epub 2010/07/14.
23. Trapp BD, Stys PK. Virtual hypoxia and chronic necrosis of demyelinated axons in multiple sclerosis. Lancet Neurol. 2009;8(3):280–91. Epub 2009/02/24.
24. Zhang F, Eckman C, Younkin S, Hsiao KK, Iadecola C. Increased susceptibility to ischemic brain damage in transgenic mice overexpressing the amyloid precursor protein. J Neurosci: Off J Soc Neurosci. 1997;17(20):7655–61. Epub 1997/10/07.
25. Koike MA, Green KN, Blurton-Jones M, Laferla FM. Oligemic hypoperfusion differentially affects tau and amyloid-{beta}. Am J Pathol. 2010;177(1):300–10. Epub 2010/05/18.
26. Gorelick PB, Nyenhuis D. Understanding and treating vascular cognitive impairment. Continuum (Minneap Minn). 2013;19(2 Dementia):425–37. Epub 2013/04/06.
27. Farooq MU, Gorelick PB. Vascular cognitive impairment. Current atherosclerosis reports. 2013;15(6):330. Epub 2013/04/25.
28. Smith E. Vascular cognitive impairment. Continuum (Minneap Minn). 2016;22(2 Dementia):490–509. Epub 2016/04/05.
29. Leys D, Erkinjuntti T, Desmond DW, Schmidt R, Englund E, Pasquier F, et al. Vascular dementia: the role of cerebral infarcts. Alzheimer Dis Assoc Disord. 1999;13(Suppl 3):S38–48. Epub 1999/12/28.
30. McKay E, Counts SE. Multi-infarct dementia: a historical perspective. Dement Geriatr Cogn Disord Extra. 2017;7(1):160–71. Epub 2017/06/20.
31. Sachdev P, Kalaria R, O'Brien J, Skoog I, Alladi S, Black SE, et al. Diagnostic criteria for vascular cognitive disorders: a VASCOG statement. Alzheimer Dis Assoc Disord. 2014;28(3):206–18. Epub 2014/03/19.
32. Roman GC, Tatemichi TK, Erkinjuntti T, Cummings JL, Masdeu JC, Garcia JH, et al. Vascular dementia: diagnostic criteria for research studies. Report of the NINDS-AIREN International Workshop. Neurology. 1993;43(2):250–60. Epub 1993/02/01.
33. Gorelick PB, Scuteri A, Black SE, Decarli C, Greenberg SM, Iadecola C, et al. Vascular contributions to cognitive impairment and dementia: a statement for healthcare professionals from the american heart association/american stroke association. Stroke. 2011;42(9):2672–713. Epub 2011/07/23.
34. Association AP. Diagnostic and statistical manual of mental disorders, 5th ed. Washington, DC; 2013.
35. Chui HC, Victoroff JI, Margolin D, Jagust W, Shankle R, Katzman R. Criteria for the diagnosis of ischemic vascular dementia proposed by the State of California Alzheimer's Disease Diagnostic and Treatment Centers. Neurology. 1992;42(3 Pt 1):473–80. Epub 1992/03/01.
36. Organization WH. The ICD-10classification of mental and behavioural disorders. Switzerland: Diagnostic criteria for research Geneva; 1993.
37. Longstreth WT Jr. Brain vascular disease overt and covert. Stroke. 2005;36(10):2062–3. Epub 2005/09/30.

38. Folstein MF, Folstein SE, McHugh PR. Mini-mental state. A practical method for grading the cognitive state of patients for the clinician. Journal of psychiatric research. 1975;12(3):189–98. Epub 1975/11/01.
39. Nasreddine ZS, Phillips NA, Bedirian V, Charbonneau S, Whitehead V, Collin I, et al. The montreal cognitive assessment, MoCA: a brief screening tool for mild cognitive impairment. J Am Geriatr Soc. 2005;53(4):695–9. Epub 2005/04/09.
40. Aprahamian I, Martinelli JE, Neri AL, Yassuda MS. The clock drawing test: a review of its accuracy in screening for dementia. Dement Neuropsycholo. 2009;3(2):74–81. Epub 2009/04/01.
41. Hachinski VC, Lassen NA, Marshall J. Multi-infarct dementia. A cause of mental deterioration in the elderly. Lancet. 1974;2(7874):207–10. Epub 1974/07/27.
42. Pendlebury ST, Cuthbertson FC, Welch SJ, Mehta Z, Rothwell PM. Underestimation of cognitive impairment by Mini-Mental State Examination versus the Montreal Cognitive Assessment in patients with transient ischemic attack and stroke: a population-based study. Stroke. 2010;41(6):1290–3. Epub 2010/04/10.
43. Wardlaw JM, Smith EE, Biessels GJ, Cordonnier C, Fazekas F, Frayne R, et al. Neuroimaging standards for research into small vessel disease and its contribution to ageing and neurodegeneration. Lancet Neurol. 2013;12(8):822–38. Epub 2013/07/23.
44. Smith EE, Schneider JA, Wardlaw JM, Greenberg SM. Cerebral microinfarcts: the invisible lesions. Lancet Neurol. 2012;11(3):272–82. Epub 2012/02/22.
45. Haridimou A, Gang Q, Werring DJ. Sporadic cerebral amyloid angiopathy revisited: recent insights into pathophysiology and clinical spectrum. J Neurol Neurosurg Psychiatry. 2012;83(2):124–37. Epub 2011/11/08.
46. Federico A, Di Donato I, Bianchi S, Di Palma C, Taglia I, Dotti MT. Hereditary cerebral small vessel diseases: a review. J Neurol Sci. 2012;322(1–2):25–30. Epub 2012/08/08.
47. Black SE. Vascular cognitive impairment: epidemiology, subtypes, diagnosis and management. J R Coll Physicians Edinb. 2011;41(1):49–56. Epub 2011/03/03.
48. Boehme AK, Esenwa C, Elkind MS. Stroke risk factors, genetics, and prevention. Circ Res. 2017;120(3):472–95. Epub 2017/02/06.
49. Pantoni L. Cerebral small vessel disease: from pathogenesis and clinical characteristics to therapeutic challenges. Lancet Neurol. 2010;9(7):689–701. Epub 2010/07/09.
50. Dufouil C, Chalmers J, Coskun O, Besancon V, Bousser MG, Guillon P, et al. Effects of blood pressure lowering on cerebral white matter hyperintensities in patients with stroke: the PROGRESS (Perindopril protection against recurrent stroke study) magnetic resonance imaging substudy. Circulation. 2005;112(11):1644–50. Epub 2005/09/08.
51. Kivipelto M, Solomon A, Ahtiluoto S, Ngandu T, Lehtisalo J, Antikainen R, et al. The Finnish geriatric intervention study to prevent cognitive impairment and disability (FINGER): study design and progress. Alzheimer's Dement J Alzheimer's Assoc. 2013;9(6):657–65. Epub 2013/01/22.

Arterial Stiffness and Aging

Sandra Morović and Vida Demarin

As the old saying goes: "We are as old as our blood vessels." We all know what aging looks like on the outside, but do we know what aging looks like on the inside? For arteries, aging means losing elasticity. Elasticity decline is defined by an increase in stiffness. Nowadays arterial stiffness is accepted to be a subclinical marker of cardiovascular and cerebrovascular disease. Arterial stiffness increases progressively with age, and its increase is in correlation with increased blood pressure. For many years now, studies have been proving an association between increased arterial stiffness and poor cognition [1]. Increased arterial stiffness is related to a greater age, male sex, African American race, lower education, higher BMI, lower HDL, higher mean arterial blood pressure, diabetes type 2, myocardial infarction, hypertension, and lower cognitive score. Some studies examined whether increased arterial stiffness is associated with a decline in specific cognitive domains. Tests were created for examining and following cognitive domains: memory, executive function, language, and visuospatial domain. Results showed that there was a great decline in visual, spatial, and language tasks in comparison to executive function or memory tasks. A study with a nine-year follow-up confirmed that higher arterial stiffness is associated with a faster rate of cognitive decline. Beyond traditional cardiovascular risk factors, such as BMI, type 2 diabetes, hypertension, and mean arterial blood pressure. Pathways explaining arterial stiffness and cognitive decline have been postulated. The first postulate says that when arteries undergo stiffness, this results in damages to pressure pulsatility. This causes hemodynamic stress in the heart and end organs such as the brain to which it is transmitted. These changes result in structural changes to cerebral blood vessels that may interfere with the transport of important nutrients to the brain and interfere with the clearance of toxic byproducts out of the brain. Also, recent

S. Morović (✉)
Juraj Dobrila University of Pula, Zagrebačka 30, HR-52100 Pula, Croatia
e-mail: sandra.morovic@poliklinika-aviva.hr

V. Demarin
International Institute for Brain Health, Ul. Grada Vukovara 271, HR-10000 Zagreb, Croatia

© Springer Nature Switzerland AG 2020
V. Demarin (ed.), *Mind and Brain*, https://doi.org/10.1007/978-3-030-38606-1_11

imaging studies link arterial stiffness to cerebrovascular disease and changes in the functioning of frontal subcortical regions of the brain. Changes like white matter hyperintensities are associated with cognitive impairment. In the end, arterial stiffness can be considered an independent predictor of cerebrovascular events and one of the important predictors of cognitive decline [1].

Consensus on Arterial Stiffness

As a non-invasive method, arterial stiffness has gained many followers. Due to the fact that, in the beginning, researchers used different methods to calculate arterial stiffness, a consensus was needed. Until then, the results were difficult to compare due to different methodologies used to assess arterial stiffness. The first guidelines for arterial stiffness measurements were introduced in 2006 in the European consensus document. The 2006 consensus document was the first publication to organize the approach to arterial stiffness, especially given the variety of approaches (tonometry, ultrasound, oscillometry, and magnetic resonance imaging) and the vascular territories studied. The main recommendations from this effort included a review of arterial stiffness physiology, measurement techniques, implications of arterial stiffness, and the effects of medications [2].

International reference standards were published in 2010, with normal and reference values for pulse wave velocity (PWV), the non-invasive gold standard for measuring arterial stiffness [3]. This huge effort included nearly 17,000 subjects (about 1,500 were normal subjects) from several cohorts and provided decade-specific values for carotid-femoral PWV in normal people, and in those with hypertension. The recommendation from this report, that clinical concern was warranted when carotid-femoral PWVs exceeded 12 m/s, was later modified by the artery research group of Europe to a value of 10 m/s [4].

The American Heart Association (AHA) scientific statement on arterial stiffness measurements in the USA, was published in 2015 [5]. These guidelines improved guidance and led to greater uniformity in clinical studies. This statement reviewed the physiology behind the arterial stiffness measurements, the value that arterial stiffness brings into the clinical field, and the various methodologies used to assess PWV, etc.

Arterial Stiffness—A Parameter of Cognitive Impairment

Vascular aging shows a strong relationship between age and changes in large artery structure and function. Several arterial parameters have been selected for clinical investigation, based on their predictive value of cardiovascular events [6]. Subclinical atherosclerosis involves intima-media thickness (IMT), plaque formation and composition, and alterations in arterial mechanisms. Studies dealing with subclinical atherosclerosis are based on B-mode ultrasound imaging [7]. Cyclic vessel distention in normal aging (i.e., beat-to-beat change in vessel diameter in response to the pulsatility of blood pressure) is thought to cause fragmentation and depletion of elastin and increased collagen deposition, resulting in an increase in IMT [8]. The relationship between carotid IMT and cognitive function has been analyzed cross-sectionally [9, 10], and longitudinally [11–13] in several studies. A significant inverse relationship between carotid IMT and cognitive function was observed in all studies; specifically, the thicker the artery, the lower the cognitive performance. This relationship was significant after controlling for age and education. Some studies further adjusted for the presence of depressive symptoms [11, 13] and/or cardiovascular risk factor levels [13]. The precise causal association of carotid IMT with vascular cognitive impairment (VCI) is uncertain. Carotid IMT can reflect either a media thickening in response to the increase in blood pressure in hypertensive patients, and intima thickening in response to atherosclerotic risk factors, or most often a combination of both. Nearly all types of vascular disease that may increase IMT may also affect cognitive function through a variety of mechanisms, directly or indirectly. Carotid atherosclerosis and IMT have been associated with cardiovascular risk factors, including metabolic, inflammatory, and dietary factors, which have also been associated with cognitive decline [13, 14]. IMT, plaque formation, and alterations in arterial mechanics are the markers of subclinical atherosclerotic changes. Carotid IMT is a sensitive marker of atherosclerosis [15]. Abnormal IMT was found to be associated with memory loss one year after stroke [16]. While the IMT measured at the time of stroke may predict cognitive decline a year later, this is not true with regard to the degree of carotid stenosis. The IMT is uniformly distributed along the arteries and its main predictors are age and blood pressure [17]. Carotid IMT was associated with an increased risk for cognitive deficits, particularly poor memory and cognitive speed in elderly women [18, 19]. In clinical practice, the measurement of IMT is relatively simple, safe, inexpensive, precise, and reproducible, and may help to monitor the response of atherosclerosis to treatment.

Recent data have described an association between high pulse pressure (PP) and Alzheimer's disease (AD) [19, 20]. PP displays a linear increase with age and represents an index of vascular aging. Alterations in pulse wave velocity (PWV) and wave reflection timing associated with age-related arterial stiffening are responsible for the increased systolic and decreased diastolic pressure (i.e., increased PP) characteristic of vascular aging. PP elevation is associated with an increased risk of AD, independent of clinical stroke, and hypertension, even in older patients [20]. Pulse pressure elevation is associated with increased cerebrospinal fluid levels of p-tau and decreased A1-42 in cognitively normal older adults, suggesting that pulsatile hemodynamics may be related to amyloidosis and tau-related neurodegeneration. The relationship between PP and cerebrospinal fluid biomarkers is age-dependent and observed only in participants in the fifth and sixth decades of life [21].

Arterial stiffness is a clinical indicator of high PP. One hypothesis is that functional changes in the arterial system may be involved in the pathogenesis of dementia. Specific measures of large artery structure and function, such as aortofemoral or brachial-ankle PWV, can be regarded as useful markers of subclinical vascular disease and are reliable predictors of vascular accidents such as stroke. Pulse wave analysis measures arterial stiffness between two points of the arterial tree [22].

Ultrasound technology is used to identify an approaching arterial pulse and measure the velocity of a pulse wave. This method is considered to represent a gold standard in arterial stiffness measurement [4]. It allows for the detection of pathology prior to the appearance of morphological changes in the vasculature.

Arterial stiffness has also been shown to be associated with cognitive changes [23]. PWV is associated with the cognitive deficit and with greater personal dependency, independently of major modifiable cerebrovascular risk factors. PWV is also a strong predictor of cognitive impairment, independent of age, gender, education, and traditional cardiovascular risk factors. Depending on further studies, PWV may be shown to be valuable as a marker of progression from MCI to clinical dementia [24].

The most simple, non-invasive, robust, and reproducible method with which to determine aortic stiffness is the measurement of carotid-femoral PWV, using the foot-to-foot velocity method from various waveforms (pressure, Doppler, distention) [6]. The analysis of aortic pressure waveform allows the calculation of central systolic blood pressure and PP, which are influenced by aortic stiffness and the geometry and vasomotor tone of small arteries. Central systolic blood pressure and PP can be estimated noninvasively either from the radial artery waveform, using a transfer function, or from the common carotid waveform. The aortic PWV, the central systolic blood pressure, and the PP each predict cardiovascular events independent of classic cardiovascular risk factors.

Carotid-femoral PWV, the "gold standard" for evaluating arterial stiffness [6], was higher in any group of cognitively impaired subjects with or without dementia [25]. An inverse relationship between PWV and cognitive performance was reported cross-sectionally. Carotid-femoral PWV was also associated prospectively with cognitive decline before dementia in studies using a cognitive screening test and more specific tests for verbal learning, delayed recall, and nonverbal memory. These relationships remained significant after controlling for age, gender, education, and blood pressure levels. Other studies reported a significant positive relationship between arterial stiffness and volume or localization of white matter lesions, a known factor known to predispose for dementia on neuroimaging [26].

Several pathways may link aortic stiffness to microvascular brain damage. They include endothelial dysfunction and oxidative stress, a mutually reinforcing remodeling of large and small vessels (i.e., large/small artery cross-talk), and exposure of small vessels to the high-pressure fluctuations of the cerebral circulation, which is perfused at high-volume flow throughout systole and diastole, with very low vascular resistance. Additionally, stiffer large arteries are associated with increased left ventricular mass. Of note, left ventricular remodeling and hypertrophy have been associated with higher frequency and severity of subclinical brain damage. Recently, higher left ventricular mass in older people has been independently associated with a two-fold higher likelihood of having dementia, independent of blood pressure levels [26].

The mechanistic pathways linking microvascular brain damage to carotid IMT, aortic stiffness, and small artery remodeling are complementary. Aortic stiffness predicts cardiovascular events independent of carotid IMT [27]. A large/small artery cross-talk has already been described in hypertensive patients [28]. These data suggest that the non-invasive investigation of large and small arteries could demonstrate additional and independent predictive values for VCI and dementia. In addition, such a non-invasive investigation could help in determining the relative weight of each arterial parameter in contributing to all types of dementia (from vascular dementia to Alzheimer's disease) in the general population [28].

Flow-mediated dilation, the ability of the vasculature to adjust flow in response to endogenous or exogenous stimuli, provides another measure of vascular function distinct from arterial stiffness. Reduced flow-mediated dilation is associated with increased cardiovascular risk and is a surrogate for vascular dysfunction [22, 29, 30]. Studies in small samples have reported an association of impaired vascular function with subclinical markers of cerebral dysfunction including white matter hyperintensities [31] and cognitive decline [32].

Elevated central arterial stiffness and pulsatility lead to distal cerebral microvascular damage and brain atrophy, which manifests in subclinical cognitive dysfunction.

Increased aortic stiffness and elevated PP may stimulate vascular hypertrophy, remodeling, or rarefaction in the microcirculation, leading to increased resistance to mean flow and impaired microvascular reserve. Endothelial function also is impaired by increased pressure pulsatility. Increased arterial stiffness is associated with abnormal blood flow patterns, which increase the transmission of pulsatile energy into the microcirculation, damaging it. These abnormal physical forces in the arteries also trigger atherogenic, hypertrophic, and inflammatory responses, which may contribute to multisystem end-organ damage. In the brain, pathologic studies have demonstrated arteriosclerotic changes in small vessels in brain regions with white matter hyperintensity [33].

References

1. Iulita MF, Noriega de la Colina, A, Girouard, H. Arterial stiffness, cognitive impairment and dementia: confounding factor or real risk? J. Neurochem. 2018;144:527–48. https://doi.org/10.1111/jnc.14235.
2. Laurent S, Cockcroft J, Van Bortel L, Boutouyrie P, Giannattasio C, Hayoz D, Pannier B, Vlachopoulos C, Wilkinson I, Struijker-Boudier H. Expert consensus document on arterial stiffness: methodological issues and clinical applications. Eur Heart J. 2006;27:2588–605.
3. Reference values for arterial stiffness collaboration: determinants of pulse wave velocity in healthy people and in the presence of cardiovascular risk factors: 'establishing normal and reference values'. Eur Heart J 2010;31:2338–50.
4. Van Bortel LM, Laurent S, Boutouyrie P, et al. Expert consensus document on the measurement of aortic stiffness in daily practice using carotid-femoral pulse wave velocity. J Hypertens. 2012;30:445–8.

5. Townsend RR, Wilkinson IB, Schiffrin EL, et al. Recommendations for improving and standardizing vascular research on arterial stiffness: a scientific statement from the American Heart Association. Hypertension. 2015;66:698–722.
6. Laurent S, Cockcroft J, Van Bortel L, et al. European network for non-invasive investigation of large, arteries. Expert consensus document on arterial stiffness: methodological issues and clinical applications. Eur Heart J 2006;27:2588–605.
7. Pignoli P, Tremoli E, Poli A, et al. Intimal plus medial thickness of the arterial wall: a direct measurement with ultrasound imaging. Circulation. 1986;74:1399–406.
8. Kohn JC, Lampi MC, Reinhart-King CA. Age-related vascular stiffening: causes and consequences. Front Genet. 2015;6:112. Published 2015 Mar 30. https://doi.org/10.3389/fgene.2015.00112.
9. Morović S, Jurasić MJ, Martinić Popović I, et al. Vascular characteristics of patients with dementia. J Neurol Sci. 2009;283:41–3.
10. Muller M, Grobbee DE, Aleman A, et al. Cardiovascular disease and cognitive performance in middle-aged and elderly men. Atherosclerosis. 2007;190:143–51.
11. Komulainen P, Kivipelto M, Lakka TA, et. Al. Carotid intima-media thickness and cognitive function in elderly women: a population-based study. Neuroepidemiology 2007;28:207–13.
12. Silvestrini M, Gobbi B, Pasqualetti P, et al. Carotid atherosclerosis and cognitive decline in patients with Alzheimer's disease. Neurobiol Aging. 2009;30:1177–83.
13. Moon JH, Lim S, Han JW, et al. Carotid intima-media thickness is associated with the progression of cognitive impairment in older adults. Stroke. 2015;46:1024–30. https://doi.org/10.1161/STROKEAHA.114.008170.).
14. Jurašić MJ, Lovrenčić Huzjan A, Bedeković Roje M, et al. How to monitor vascular aging with an ultrasound. J Neurol Sci. 2009;1–2:139–42.
15. Rundek T, Demarin V. Carotid intima-media thickness (IMT): a surrogate marker of atherosclerosis. Acta Clin Croat. 2006;45:45–51.
16. Talelli P, Ellul J, Terzis G, et al. Common carotid artery intima media thickness and post-stroke cognitive impairment. J Neurol Sci. 2004;223:129–34.
17. Spence JD. Carotid intima-media thickness and cognitive decline: what does it mean for prevention of dementia? J Neurolo Sci. 2004;223:103–5.
18. Komulainen P, Kivipelto M, Lakka TA, et al. Carotid intima-media thickness and cognitive function in elderly women: a population-based study. Neuroepidemiology. 2007;28:207–13.
19. Demarin V, Bašić Kes V, Morović S, et al. Neurosonology: a means of evaluating normal aging versus dementia. Aging Health. 2008;4:529–34.
20. Nation DA, Edmonds EC, Bangen KJ, et al. Pulse pressure in relation to tau-mediated neurodegeneration, cerebral amyloidosis, and progression to dementia in very old adults. JAMA Neurol. 2015;72(5):546–53. https://doi.org/10.1001/jamaneurol.2014.4477.
21. Nation DA, Edland SD, Bondi MW, et al. Pulse pressure is associated with Alzheimer biomarkers in cognitively normal older adults. Neurology. 2013;81:2024–47.
22. Demarin V, Morović S. Ultrasound subclinical markers in assessing vascular changes in cognitive decline and dementia. JAD. 2014;42:259–66.
23. Matsuoka O, Otsuka K, Murakami S, et al. Arterial stiffness independently predicts cardiovascular events in an elderly community—longitudinal investigation for the longevity and aging in Hokkaido county (LILAC) study. Biomed Pharmacother 59;2005:(1)S40–4.
24. Singer J, Trollor JN, Baune BT, et al. Arterial stiffness, the brain and cognition: a systematic review. Ageing Res Rev. 2014;15:16–27.
25. Hanon O, Haulon S, Lenoir H, et al. Relationship between arterial stiffness and conitive function in elderly subjects with complaints of memory loss. Stroke. 2005;36:2193–7.
26. Mancia G, Fagard R, Narkiewicz K, et al. 2013 ESH/ESC guidelines for the management of arterial hypertension: the task force for the management of arterial hypertension of the European Society of Hypertension (ESH) and of the European Society of Cardiology (ESC). J Hypertens. 2013;31:1281–357.

27. Mattace-Raso FU, van der Cammen TJ, Hofman A, et al. Arterial stiffness and risk of coronary heart disease and stroke: The Rotterdam Study. Circulation. 2006;113:657–63.
28. Laurent S, Briet M, Boutouyrie P. Large and small artery cross-talk and recent morbidity-mortality trials in hypertension. Hypertension. 2009;54:388–92.
29. Kohn JC, Lampi MC, Reinhart-King CA. Age-related vascular stiffening: causes and consequences. Front Genet. 2015;6:112. https://doi.org/10.3389/fgene.2015.00112.
30. Yeboah J, Crouse JR, Hsu FC, et al. Brachial flow-mediated dilation predicts incident cardiovascular events in older adults: the cardiovascular health study. Circulation. 2007;115:2390–7.
31. Burger R, Toyuz R. Cellular biomarkers of endothelial health: microparticles, endothelial progenitor cells, and circulating endothelial cells. J Am Soc Hypertens. 2012;6:85–99.
32. Gonzales MM, Tarumi T, Tanaka H, et al. Functional imaging of working memory and peripheral endothelial function in middle-aged adults. Brain Cogn. 2010;73:146–51.
33. Tsao CW, Seshadri S, Beiser AS, et al. Relations of arterial stiffness and endothelial function of brain aging in the community. Neurology. 2013;81:984–91.

Zeitfracht Medien GmbH
Ferdinand-Jühlke-Straße 7
99095 Erfurt, Deutschland
produktsicherheit@kolibri360.de